IMA MONOGRAPH SERIES

IMA MONOGRAPH SERIES

1. **H. R. Pitt**
   Measure and integration for use

2. **James Lighthill**
   An informal introduction to theoretical fluid mechanics

3. **Matiur Rahman**
   *Water waves: relating modern theory to advanced engineering
   applications*

4. **Kenneth W. Kemp**
   The efficient use of quality control data

The efficient use of quality control data

Kenneth W. Kemp

Emeritus Professor, University of Wales

CLARENDON PRESS • OXFORD
2001

OXFORD

UNIVERSITY PRESS

Great Clarendon Street, Oxford OX2 6DP

Oxford University Press is a department of the University of Oxford.
It furthers the University's objective of excellence in research, scholarship,
and education by publishing worldwide in

Oxford New York

Athens Auckland Bangkok Bogotá Buenos Aires Cape Town
Chennai Dar es Salaam Delhi Florence Hong Kong Istanbul Karachi
Kolkata Kuala Lumpur Madrid Melbourne Mexico City Mumbai
Nairobi Paris São Paulo Shanghai Singapore Taipei Tokyo Toronto Warsaw

with associated companies in Berlin Ibadan

Oxford is a registered trade mark of Oxford University Press
in the UK and in certain other countries

Published in the United States
by Oxford University Press Inc., New York

British Library Cataloguing in Publication Data

Library of Congress Cataloging in Publication Data
Kemp, Kenneth W. (Kenneth Walter), 1925–
The efficient use of quality control data / Kenneth W. Kemp.
(IMA monograph series; 4)
Includes bibliographical references and index.
1. Quality control. 2. Quality assurance. I. Title. II. Series.
TS156.K418 2001 658.5′62–dc21 00-067604
ISBN 0-19-853674-7 (Hbk.)

Typeset by Newgen Imaging Systems (P) Ltd., Chennai, India
Printed in Great Britain on acid-free paper by
T J International Ltd, Padstow, Cornwall

Preface

The application of statistical method in practice is an art as well as the application of a set of mathematical disciplines. It can certainly be described as the art of approximation. In the practical application of decision rules to control the quality of a product or constancy in the level of a certain clinical test, we may need to consider the accuracy with which the characteristics of schemes need to be computed on the one hand, and the simplicity of its design and operation on the other. For the control of clinical testing ethical constraints can substantially limit the design of the perfect scheme. Generally speaking, one finds such constraints frequently limit the design of clinical investigations.

Much of the material in the monograph is concerned with the control of continuously operating industrial processes and the use of statistical decision rules in clinical assays. For both, attention is confined to situations where once the process or testing procedure is set in motion the target in control levels will usually be maintained for long periods of time. An industrial example is a chemical engineering production process to spin synthetic yarns for the textile industry.

It is evident, from media coverage for example, that the inevitability of risk in many aspects of life is not in general appreciated by the populace as a whole. This is particularly so with regard to expectations from modern medical treatment. Experience reveals that technical staff responsible for routine clinical testing are also not particularly well versed in contemporary risk theory. International conferences run, for example, by the Tenovus Institute in Cardiff revealed this to be an acute problem in the area of assay testing. Despite considerable advances in many aspects of industrial quality control this is still the case in some industrial organizations, particularly those operating on a small scale. This monograph accordingly is meant to assist those responsible for controlling laboratory testing. It discusses the inevitability of wrong decisions and control of the frequency with which they will occur. It describes the use of statistical decision rules and the consequence of their use both in routine clinical testing and certain types of industrial processes. It is written for clinical technicians and quality control engineers with a reasonable knowledge of mathematics who need to familiarize themselves only with those aspects of statistical method relevant to keeping clinical test levels or production processes at specified target levels. Readers with a limited formal mathematical background will find that the text contains many examples which have been selected to illustrate principles of control in an arithmetic fashion.

A major objective is to illustrate the considerable improvements in the characteristics of control rules that can be achieved by appropriate choice of parameters, test statistics and methods of control which fully utilise information in test data. A further objective is to illustrate that control rules can be found which are close to the best which can be devised. We do so by defining criteria which can be used to identify best schemes and illustrate that the definition of best schemes depends upon the field of application. The monograph is in three distinct parts.

Part I consists of Chapters 1 to 4. It presents only those concepts of statistical theory needed to design decision rules on which quantitative control procedures are based. Chapter 1 discusses a number of general aspects of quality control, briefly indicating the progress which has been made in particular over the last fifty years. It includes a description of a quality control problem in which the author became involved many years ago. The reason for doing so is to illustrate that many different disciplines can be required in order to set up and implement quality control schemes whether they be in industry or medicine. Setting up such schemes will usually require a much wider range of investigation and solution of technical problems than the mere design of a control chart. They will almost certainly invoke the application of a number of different aspects of statistical methodology.

It discusses those basic properties of functions of test data relevant to the design of control rules. It suggests a practical interpretation of acceptable and rejectable quality levels and highlights the consequences of using statistical decision rules.

Part II consists of Chapters 5 and 6. Using the definitions described in Part I it uses simply designed control charts together with the concepts of acceptable and rejectable levels to form a simple model on which to establish principles and criteria for the design and operation of statistical control procedures.

The criteria developed in Part II lead to the definition of envelopes in Part III which permit comparisons between different methods of control and the identification of best schemes.

Part III is Chapter 7. It describes the use of cumulative sums which aim to make full use of all of the information contained in sampled data which relates to the value of the population parameter being controlled. The technique relates to clinical control procedures which use control pools to ensure testing is in control, and industrial processes which once set in motion can be expected to run at an in control quality level over long periods of time. The use of a control pool large enough to last a considerable time, or industrial processes of the kind just described, permits us to relate the data of successive control samples one to another in a sequential fashion. This approach has been developed in recent years and is particularly appropriate in clinical testing where it is being increasingly used. Part III discusses the cumulation of successive results in detail. It indicates that some charts with warning and action lines are in fact cumulative sum charts with crude scoring systems which accordingly do not use the sampled information efficiently. Tables given in this part of the monograph illustrate that the method of selectively calculating cumulative sums leads to schemes which make very efficient use of sampled data. The definition of envelopes leads to the identification of decision rules which are the best that can be achieved. They also lead to the conclusion that close to best schemes can be achieved with the simultaneous use of just two cumulative sum decision rules.

The mathematics needed to determine the properties of the schemes described in this third part of the monograph is unfortunately more complicated than that required in Part I and II. However, it ultimately leads to the formulation of simple numerical expressions such as eqns (7.68) to (7.75) which are easy to use in the design of practical control rules. These expressions as far as I am aware are new, they give very close approximations to true run length values. Equation (7.97) is also new. Its use with central reference values leads to the straight forward design of snub-nosed V-masks or equivalently the simultaneous use of two cumulative sum schemes.

An important aspect of clinical testing in which the author became involved was the control of test variability. In many clinical applications this may only be achieved with duplicate or at most triplicate testing. This aspect of clinical control is treated in Section 7.20. An additional objective here is to emphasize the generality of the reasoning leading to and including eqn (7.69). It illustrates that the equations and methods used to formulate this expression can be employed to calculate the run lengths of cumulative sum schemes for cumulations of statistics which are not normally distributed. With the computing facilities now available it also illustrates the use of properly designed simulation procedures to determine the values of characteristics such as run lengths when the distributions of cumulated statistics have unpleasant features like singularities!

The monograph also indicates aspects of the use of the central limit theorem in the design and operation of control procedures. Examples 5.3 to 5.6, for instance, illustrate the simplicity with which parameter values can be determined with its use. Control rules which emerge can be used in practical situations where there is a certain ambivalence with regard to sample size or the exact probabilities of right and wrong decisions. On the other hand, as for instance in Example 6.5, the theorem can be used to indicate where to look for parameter values to control process or testing standard deviations. The example demonstrates that it is relatively easy to adjust values given using the theorem to those of the actual distribution of the sample statistic.

Finally, I think it necessary to say that the reader will be subjected to a number of personal observations about some aspects of statistical quality control. They are remarks which I believe still need to be made. They result from many years experience in the design and implementation in practice of control schemes and the theoretical development of new techniques for the control of continuously operating industrial plant and latterly the control of clinical testing levels.

The monograph only draws attention to those statistical criteria relevant to the efficient use of test data in the design of control rules. Their formal derivation can be found in standard works which are far too many to mention here except to acknowledge the debt owed by the author to them The bibliography in the main contains only references relevant to material in this monograph. It does, however, contain a list of research papers concerned with matters discussed in the text, in particular those relating to various aspects of clinical assays which have received much attention in recent years.

In conclusion I would like to acknowledge the considerable contributions made to the issues discussed in this monograph by Keith Griffiths, the Director of Tenovus, and his colleague Doug Wilson, together with my indebtedness to my colleagues Jim Rowlands and Barry Nix. I am also particularly indebted to a former doctoral student Malihe Akhavan-Abdollahian for carrying out carefully designed and demanding computer simulations relating to the conclusions discussed at the end of Chapter 7. I am also grateful to Professor John Blake for a number of suggestions which have substantially improved the presentation of this monograph.

July 2000   K.W.K.

Contents

Part I

Statistical concepts

1
Some aspects of statistical quality control

1.1 Historical note

The increasing use of statistical methods to monitor the quality of manufactured goods could give the impression that quality control is a contemporary innovation. Pre-occupations of the media with the safety of preservatives, foods, drugs, and the quality of many domestic items probably re-enforces this view. Controlling output quality, however, is as old as procedures used to produce goods for general consumption. For example, inspection procedures to ensure high quality glass and porcelain were used long ago. A recognition that statistical procedures, in particular small sample theory, could be used in the control of quality did not occur until comparatively recently The first recorded application of such principles to quality control appears to be contained in a Bell Telephone memorandum by W. A. Shewhart (1924). In the years that followed, quite a few publications appeared describing these charts, as did tables to facilitate their application. It is surprising that their use to monitor the output of large-scale industrial processes was much less than would have been expected.

Significant developments in both the theoretical and practical aspects of statistical quality control did not occur until the onset of World War II. Whilst it lasted and even after it ended, considerable advances in production methods and technology occurred. These were due to the demands of modern warfare during the war years and a developing society thereafter. Many of the methods of statistics then began to be used in both research and development. This led to a recognition of its methods as an essential tool for industrial and technological development. It was accompanied by an increasing awareness of the need to apply statistical method to a whole variety of manufacturing procedures both old and new. Its use has now become a matter of course in very many light and heavy chemical engineering processes for example. A significant factor in the formulation of control rules has been the increasing use of small sample theory to monitor test levels and test error for a whole variety of clinical procedures used in the diagnosis of disease. Their very nature imposes limitations on the amount of testing it is possible to carry out; sometimes we find that as few as two tests can be taken on each sampling occasion. When this, and similarly small samples are the only possibility, the information contained in the test data needs to be utilized as efficiently as possible. The search for control procedures which do so requires the formulation of criteria with which to compare the properties of different methods of control. Their formulation has led to the concept of best schemes and the subsequent identification of control schemes which for all practical purposes can be regarded as the best which can be found.

1.2 Control

Establishing and operating procedures for the control of many industrial processes can be a complicated activity requiring the cooperation of production, management, technical, and marketing expertise. It is important for the statistician involved to recognize this. The introduction of sampling and statistical methods of control is an essential part of maintaining specified levels of output when systematic and random elements in quality variation are unavoidable. Controlling the quality of material produced usually embraces much more than the effective operation of a number of control charts. We can illustrate this observation by taking an example. Let us take the case of a chemical engineering plant which is to be set up to market a new product. Suppose that in common with many such plants the methods of production are complex both from an engineering and chemical point of view. In these circumstances it is often necessary to build a main manufacturing plant with attached research and development units. Within this structure a responsibility of the research unit is frequently to identify causes of changes in quality level and variability together with devising ways to rectify them. Experimentation needed to do so will usually involve several scientific disciplines and statistical methods. Procedures and instrumentation, devised to control quality, which work in the environment of the research laboratory frequently do not do so in the less ideal surroundings of a mass production plant. It is normally the task of the development unit to get round these problems, converting research recommendations into practical tools. It is not unusual to find that considerable effort and skill is needed to do so. This is particularly the case when there is a need to produce test procedures which can be easily understood and applied by production personnel. Instrumentation, when required has to be sufficiently robust to withstand the vicissitudes of the shop floor! When there is a basic component of random variability in the process it is clear that many contemporary statistical techniques will be required in this work. They will play a significant role both in the research and development needed to produce a quality product.

For the application of statistical method to be effective in practice it is essential to present its concepts in ways which make sense to the non-statisticians involved in a project, whether it be the development of a routine testing procedure or a fundamental research investigation. In medicine, for example, clinicians and laboratory personnel need to be clear about the nature of conclusions which can be drawn from a statistically based study. In industry, research, production, and marketing people need to be aware of the consequences of using statistical decision rules. With regard to statistical quality control we can illustrate these observations with a particular example. Suppose a manufacturing process is such that the production of a proportion p of defective items is unavoidable. Suppose also it is only possible to test a sample of items in any one production batch. In order to market the product it is necessary to control the value of p. To do this we need to define certain parameters beginning by identifying two values of p, namely p_a and p_r ($p_r > p_a$). If in a particular batch $p < p_a$ then it can be supplied to the customer. If, however, the proportion of defective items is $> p_r$ then the batch is rejected; p_a is called acceptable quality level (AQL) and p_r is called rejectable quality level (RQL). With these quality levels we can develop statistical rules to control the proportion of defective items which will be supplied to customers. We shall see that this approach is equivalent to forming a simple mathematical model of the problem of controlling the value of p. It is important for the statistician involved in such work, to recognize that devising methods to produce a quality product by concentrating on only two values of p, could be regarded by production management as an over-simplification of the problem. Notwithstanding this observation, it is equally important to recognize that the formal treatment of simplified models usually leads to the formulation of methods which can be used in more sophisticated situations.

Let us explore the example of defining just two values of p a little further. Suppose that quality control charts are to be introduced into a large-scale production unit to monitor p and that a statistician is asked to design and implement their application. What difficulties might he experience? He may find it difficult to justify to production and marketing people that a consequence of using his methods will result in the occasional rejection of batches of a perfectly acceptable product. The idea that customers will also from time to time inevitably receive substandard material might cause problems.

Control models must be seen to be realistically formulated. The criteria for decision rules based on them must also be relevant to specific fields of application. It is important to have sensible modelling on the one hand and develop appropriate decision criteria on the other. There has been a tendency to ignore these two considerations both in the theoretical and practical application of quality control. An important example is the application of statistical control to clinical testing, which we shall discuss shortly. Before doing so let us consider a little further the control of the proportion of defective items p, in particular the specification of just two values for p. Why might schemes designed to decide whether or not p is less than or equal to p_a, or greater or equal to p_r, be regarded by management with some unease? Are control rules based on them likely to accord with the objectives of manufacturer or customer? When procedures based on just two values of p were first formulated a number of years ago, the range of values of p between p_a and p_r was called the indifference quality region. This terminology could imply an ambivalence about batches with a proportion of defective items between p_a and p_r. Obtaining values to assign to p_a and p_r from either manufacturer or customers is usually not an easy matter. The former wants to sell as much of his product as he can and the latter prefers not to take any defective material. In reality neither producer or customer is, in fact, 'indifferent' about any value of p, the more so if it lies between p_a and p_r. In truth separating p into two ranges, $p < p_a$ and $p > p_r$ with a gap between p_a and p_r is a convenient mathematical device. It permits the statistician to develop criteria and principles on which to design methods of control. It is important for an inexperienced statistician to bear this very much in mind when trying to determine what values of p_a and p_r are appropriate to a particular situation As a young statistician the author learnt this lesson the hard way! To ask a manufacturer or customer 'to what values of p are you indifferent' can evoke some extremely embarrassing not to say rude replies! In the manufacturer's mind a single value of p_a does not exist, in the sense that a batch with p marginally greater than p_a cannot be sold. Equally p_r does not exist on the basis of indifference to batches with p marginally less than p_r and rejecting those with p marginally greater than p_r. Clearly, the consequences of producing a proportion p of defective items change, as the value of p changes. It is often very difficult to quantify these changes. This is particularly so when the material is sold on to other manufacturers for different processing or different end uses. As far as customers are concerned it can also be difficult to attach a cost to receiving differing proportions of defective goods. We can appreciate how difficulties arise if we consider for a moment the cost C of selling substandard material as one criterion to control p. The value of C will change as p changes; we can indicate the relationship between C and p by writing C as $C(p)$ and calling it a cost function. It is most unlikely that $C(p)$ will be the same for both manufacturer and customer. Accordingly it is realistic to anticipate at least two cost functions $C_1(p)$ and $C_2(p)$ will be needed. The first being the cost function of the manufacturer and the second that of the customer. Both will be related to each other since a manufacturer's cost function must obviously take into account the costs to his customers of selling material they may not be able to use. Evidently $C_1(p)$ should include $C_2(p)$ as a constituent element so that if $C_2(p)$ changes so will $C_1(p)$. Is it likely that the costs of receiving unacceptable material are the same for all customers? Put formally, is an assumption of a common cost function for all customers realistic? If, as with a textile yarn the material produced by a particular manufacturer is used by different customers to produce different goods we can expect a whole range of different cost functions, $C_2(p)$, all of which affect $C_1(p)$. From these almost self-evident remarks it is clear that the subject of formulating cost functions appropriate to a given set of circumstances may not be very straightforward.

Such considerations, however, should not lead to the impression that it is not possible to formulate criteria for the design of effective methods of control. Their purpose is to indicate the need for care in the design, presentation, and implementation of statistical quality control and to illustrate that introducing its methods may not be as straightforward as is often thought! In due course we shall need to consider the features of different decision rules over ranges of values of p. When doing so we use a criterion called an operating characteristic. We can think of this as a graph of a particular characteristic of a method plotted against p. We shall see that we can find different schemes with common features when $p = p_a$ and $p = p_r$, but different features for other p values. The choice of which scheme to use may well be based upon which one has the best operating characteristic or the operating characteristic which most closely reflects considerations of the cost of wrong decisions. The complex nature of identifying quality objectives and quantifying the risks of inevitably reaching wrong decisions from time to time has led many organizations to set up specification committees to carry out these tasks. Their membership usually includes production, research, and marketing personnel as well as statisticians. The expertise of such a committee and use of the operating characteristics of different control rules leads to a rational procedure for the determination of values for p_a and p_r. Essentially we treat p_a and p_r as parameters defined by the statistician to establish theoretical foundations on which methods of statistical control can be based. For given values of p_a and p_r we can calculate the operating characteristics of a whole variety of different control rules. These tell us how well particular schemes perform over all values of p. By considering criteria like cost functions we can roughly determine the shape of the operating characteristic needed. We can compare this with those of the methods of control available and identify the type of scheme we should use. This approach when explained carefully, goes a long way towards removing the difficulties industrialists and the like have with concepts such as acceptable and rejectable quality levels. It helps them to look upon p_a, for example, as a value which p takes in the region of values of p which represent an in control situation. In a sense it is a summarizing measure of them. Likewise, p_r summarizes values of p which are unacceptable. In my own experience it is not difficult to get people to give values to p_a and p_r when they are so described. This is particularly so when we explain that the scheme based on these two values will be judged upon its features over a wide range of values of p! It is very much in the interests of statisticians involved in quality control to recognize the value of a consulting committee whose members have substantial experience in the manufacture and marketing of the product being controlled. In other fields of application he should seek the guidance of a similar group of experts. In clinical testing, for example, one requires the expertise of clinicians, biochemists, and laboratory personnel to assess the consequences of changes in the levels and accuracy of testing used in disease diagnosis and treatment.

As the methods of statistics became more widespread in the 1940s there was a tendency to apply standard techniques to a variety of new fields of investigation somewhat uncritically. An example was the use of methods of experimental design, developed to conduct agricultural field trials, to study industrial production processes. These designs were formulated to investigate factors such as the crop yield obtained using different kinds and levels of fertilizer. One objective was to minimize the influence of fertility gradients of fields used in the investigations. They were also based on simple linear modelling. The relevance of such designs to the examination of the quality output of a bank of knitting machines in a textile mill, for example, is not immediately obvious! It is reasonable to suppose that procedures appropriate to agricultural field trials may need some modification before applying them to industrial production problems or the design of clinical trials! As time progressed, procedures appropriate to industrial experimentation were developed, for example, by Box, Deming, and Taguchi. A similar approach is required with regard to clinical research. One concern of this monograph is the application of the methods of statistical control to clinical testing. It is unrealistic to assume that the criteria used to control industrial processes are automatically relevant to medical testing. The technical facilities available on the factory floor are not usually as sophisticated as those in clinical laboratories. Some control procedures in industry are used for no better reason than that they are easy to operate. There will be circumstances in medical testing where this consideration is also valid. Simple methods of testing usually require a lager number of tests than those needed by more efficient ones. For many industrial processes the cost of extra sampling to achieve simplified testing can be a very secondary consideration. For much clinical testing this is not the case.

1.3 Industrial example

Let us consider a chemical engineering quality control problem in which the author was one of the statisticians involved. It was an investigation undertaken a number of years ago but it illustrates the wide variety of technical and statistical skills which can be needed in the formulation of a control scheme.

(i) *The fault*
A particular synthetic yarn when knitted into panels suitable for hosiery and dyed sometimes had an unacceptable stripy appearance called ringiness. Its cause was unknown. Dyed finished products such as fully fashioned stockings were unsaleable.

(ii) *Salient features*
At the time, a large-scale manufacturing unit had been set up which was selling its products to a substantial market. Although this was so, the process was complex and not well understood both from an engineering and chemical point of view. In addition the chemistry and technique of dyeing synthetic fabric was also at an early stage of development. It was thought that ringiness could be a dyeing problem, an unknown chemical phenomenon which occurred somewhere in the spinning process, or an engineering fault. It soon became clear that considerable effort would probably be needed to identify its cause. In the meantime sufficient non-ringy yarn was being produced to ensure commercial viability, provided quality screening procedures were introduced to limit the amount of ringy yarn sold to hosiery outlets. The statisticians first task was to set this up. It turned out that this, in the end, required more statistical input than the mere design and introduction of a few control charts. The statisticians' second task was to become involved in the design of investigations to reduce the amount of ringy yarn being produced

(iii) *Cost and test considerations*
The manufacture of the fibre from polymer chips to the finished undyed yarn was expensive. The cost consequences of reaching wrong conclusions about ringy yarn were complicated by the following factors;

(a) the yarn could only be tested at the final stage of manufacture when it had been wound on to bobbins and was ready for despatch to the trade;

(b) it was only possible to test short lengths of yarn on the outside of a bobbin;

(c) only a very small proportion of the total production could be tested;

(d) many kilometres of yarn were wound on a single bobbin so that a large number of articles such as stockings could be produced from just one bobbin;

(e) it was perfectly possible for yarn to be ringy in the middle of a bobbin and not on the outside and vice versa.

In addition to these considerations a sequence of different and separate procedures had to be applied to the yarn after it had been spun. Each of these additional production stages was also expensive. Thus, unnecessary manufacturing costs could be avoided by identifying faulty yarn production as early as possible in the production process. For many chemical engineering plants of this kind this is more easily said than done. Even after the cause of ringiness had been found and laboratory instrumentation developed to detect it, we found that a great deal of development work was needed to produce instruments which were sufficiently robust for use in the factory

All of the customers for which ringiness constituted defective yarn were hosiery manufacturers. Two types of knitting machine existed. One produced fully fashioned stockings, the other steam-set stockings. There were in consequence at least two different hosiery manufacturers to whom yarn was sold, those with fully fashioned machines only and those who manufactured only steam-set stockings. A third group were knitters who had a mixture of both kinds of machine. The cost of producing fully fashioned stockings was greater than that of the steam-set method. The evidence that ringy yarn had been used in producing a batch of hosiery became apparent only after the final stage of knitting, setting, and dyeing. Obviously from these facts, the cost of receiving ringy yarn differed from one hosiery manufacturer to another. In addition, some had their own dyeing plant whilst others did not, which was another reason for cost differences. A further difficulty in quantifying retail costs is fashion. At one time women preferred fully fashioned stockings but then the fashion changed and the seamless (steam-set) variety became popular. When this happened the price of seamless stockings did not reflect their cost of manufacture, it reflected demand whilst the fashion for them lasted. Our example shows that attaching costs to the selling and receipt of different quality levels of material may not be altogether straightforward. Clearly, an assessment of the costs to hosiery manufacturers we have just described, must be taken into account by the yarn manufacturer. His problem, as far as the costs of supplying quality continuously spun yarn to customers is concerned, has to take account of all of the different outlets to which it goes. Thus, suppose he has a batch of yarn which control testing leads him to believe it is ringy what can he do with it? Can it be diverted to an end use other than hosiery? For the yarn concerned there were a number of alternatives. It could for example be converted into synthetic wool and sold to the wool trade. The cost functions for this end use are of course very different to hosiery and equally complex. He could on the other hand divert the batch to the weaving trade where ringiness is again not a problem. If he does so then he must ensure that the defect of stripiness is not inherent in the yarn. Stripiness is a fault which can occur in synthetic yarns which is independent of ringiness and which of course generates its own cost functions.

Discussion of the complexities of formulating cost functions is in fact glossed over in many textbooks. This aspect of control often merits no more than a few sentences. It is clear from our example that for just one product there can be different sets of functions $C_1(p)$ and $C_2(p)$ according to the end use for which it is used and the nature of the particular defect. We have intimated that it can be very difficult indeed to quantify such functions. This is particularly the case with regard to clinical testing. One way of dealing with the problem is to ensure long production runs in industrial applications, and long runs in testing procedures for clinical applications when testing or production is in control. The identification of such schemes is the major topic of this monograph.

1.4 Control and removal of ringiness

In the author's experience the ringiness problem was typical of the kinds of tasks that a statistician may be expected to deal with in manufacturing industry. Hopefully it will give inexperienced readers a picture of the breadth of statistical and other expertise required to solve a specific practical problem. With this objective in mind, let us briefly consider how the cause of the defect was found and defective yarn subsequently segregated.

1. *Time taken*
It is often assumed that the design and implementation of a quality control system is a relatively trivial activity which will accordingly take little time and effort to bring about. The ringiness problem in fact took several years of full-time effort to solve and control.

2. *Expertise required*
The team of people formed to tackle the problem consisted of chemists, physicists, in particular electronics specialists, engineers, and several statisticians.

3. *Fault identification*
There were many possible factors which individually or jointly could be the cause of ringy yarn. All of these and combinations of them had to be investigated. Measurements of separate and combinations of each contained random components of variation which could not be eliminated. Application of statistical methods of experimental design was therefore necessary to link or disassociate factors which might cause ringiness.

4. *Recording equipment*

When the cause of the fault was identified electronic equipment had to be developed in the laboratory which would reliably record its occurrence.

5. *Practical tool*

When 4. had been achieved it was then necessary to develop robust equipment which could be installed on machines operating under production conditions. This had to be done because of the need to isolate ringy yarn as early as possible in the manufacturing process. The environment for sensitive equipment of this kind was far from ideal. Nevertheless, test stability had to be achieved. This particlar aspect of the project took development engineers, physicists and statisticians a considerable time to achieve.

6. *Assessment*

The cause of ringiness was found to be a short-term variation in thickness of the yarn. Instrumentation developed and subsequently used to control the fault recorded the denier (thickness) of yarn along its length. A typical trace is shown in Fig. 1.1. Measuring thickness along yarn samples and then knitting them up showed that ringiness was seen when changes in the filament width occurred within particular lengths of yarn. For ringiness to appear, a number of peaks like the two indicated in the diagram, had to occur together. Furthermore, changes in thickness had to be sustained over short lengths of yarn. A gradual increase in denier would not be seen as a ring. How were we to attach a measure to recordings like Fig 1.1, which reflected these variabilities in thickness and could easily be used in the plant? It was decided to select sections along a trace and assign a score to each one. The value given to a trace was the sum of scores for all of the sections sampled. A transparent ruler with a fixed number of sections was constructed as shown in Fig. 1.1. For reasons which need not concern us here the opaque gaps between the windows of the ruler had random widths. We found that this method of quantifying traces worked well. We also found that enumerate personnel responsible for segregating ringy yarn in the main plant could use this method of assessment with ease.

7. *Control scheme*

This scoring method was found to be both effective and practical; consideration then had to be given to the statistical quality control procedure which should be used with the measurements obtained in this way. From the nature of the continuous production process and the difficulties and expense of testing, it was clear that the proportion of yarn which could be tested was very small indeed. It was evident that information available in sampled data would need to be fully utilized, and that newly estabished sequential control schemes would have to be used. The mathematics required to obtain their properties turned out to be complicated, as was the determination of parameter values required for the design of practical schemes. With regard to the latter it became necessary to formulate easy ways to determine these.

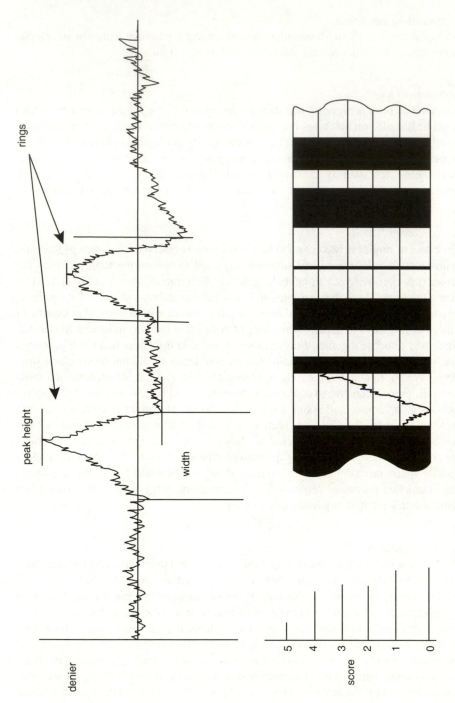

Fig. 1.1 (Above) Continuous trace of yarn denier. (Below) Section of ringiness rule.

8. *Finding the fault*
An interesting aspect of the statistician's role in technological investigation is the
ingenuity which is sometimes required of him. The present example is a case in
point. Once statistical procedures were in place to segregate defective yarn, it was
necessary to examine the production process to find its cause. Often, as in this instance,
existing statistical or mathematical techniques may not be adequate for the task, so
that new methods have to be found. In this instance the data which had to be analysed
were continuous time series similar to the one shown in Fig 1.1. We soon discovered
that these could not be modelled as systematic fluctuations in yarn thickness with
random fluctuations superimposed on them. The data suggested thickness variations
with varying periods and amplitudes with the addition of random fluctuations. This
being so, the statistician needed to suggest methods of analysis which might be used
to make sense of the traces being recorded. A wide-ranging survey of contemporary
modern methods of time-series analysis, although very interesting, proved fruitless.
There were many instruments being marketed which automatically computed full
Fourier analysis of such traces. Their use was also not appropriate for our purpose,
since they were based on assumptions about the basic structure of the data which
were not valid! After very considerable effort, looking for, and using many different
methods of analysis, it was concluded that the nature of variations in yarn thickness
was so haphazard that it was not realistic to model them mathematically. It seemed
that no appropriate technique of contemporary time-series analysis was suited to our
needs! What were we to do? How were we to obtain a method of analysis which was
quick, easy to use and interpret and importantly did not rely on assumptions about
the nature of the non random variations in the traces? Whilst pondering this problem
one day when in the library, the author was struck by a strong similarity between yarn
denier traces and a published time-series recording annual measurements of sun-spot
activity, deduced from an analysis of distances between tree rings! An article, which
appeared in an astrophysics journal in 1915 described an optical device to analyse
such data. This instrument, called an optical periodograph, was built with the help
of a very capable physicist, and found to be ideal for the analysis of data needed to
establish the cause of ringiness.

1.5 Clinical testing

A whole variety of tests carried out in hospital laboratories have a residual component
of variability which we can regard as random. Accordingly statistical procedures are
needed to ensure that test precision and accuracy are maintained at acceptable levels.
Monitoring such levels and error within a particular hospital laboratory is called
internal quality control. External quality control is the name given to procedures used
to attain conformity in these between hospitals. This is done by sending material with
known properties to different hospitals and comparing the results of their separate
analyses. The statistical aspects of the two differ considerably. In this monograph we

concentrate on the problems of internal quality control. The first published clinical application of the use of simple control charts seems to have been by Levey and Jennings in 1950. As already remarked, the assumptions and criteria on which the methods used are based need to be reviewed in new areas of application. In clinical testing this is certainly so. Testing and maintaining acceptable levels of them is often complex and costly. As methods of statistical control began to be used in this field it became increasingly clear that in many cases we would need schemes based on small samples. This implies effectively utilizing all of the information contained in the data. The mathematics of schemes which do this is more complicated than those usually employed in industry. Notwithstanding this observation, they need to be presented in a form which is easily assimilated by technical personel unfamiliar with the theoretical procedures used to formulate them. As with industrial applications it is essential to involve everyone responsible for the organization and running of routine laboratory testing. The expertise required would include,

(a) statisticians;
(b) computerization with particular reference to automatic computation and visual presentation;
(c) the technology of assays;
(d) the day-to-day in house problems of routine laboratory testing.

1.6 Efficient use of data

To introduce the concept of best tests let us consider the example of clinical laboratories where it is necessary to keep routine testing under continuous review. This is to be achieved by the control of parameters whose values have to be estimated from sampled data. A typical example would be a laboratory which daily assays a large number of patient samples where it is essential to control test level and accuracy. The method used is to monitor the mean and standard deviation of tests carried out on control material with known characteristics which are introduced at regular intervals into the testing procedure

Suppose the preparation of large quantities of control pool is cheap and that testing is inexpensive and technically straightforward. The choice of which kind of control scheme to use will then be relatively unimportant. Dominant considerations in its choice will be simplicity of operation, computerization, and interpretation. Whether or not the scheme chosen is best, in the sense we shall soon discuss, is most probably irrelevant, since we can overcome its shortcomings by merely increasing the number or size of the control samples taken from the pool. However, we cannot always use this option. The preparation of a control pool can be a technically difficult and costly procedure. In these circumstances we want the life of the pool to be as long as possible. Accordingly the number of control samples which can be introduced into routine testing at any one time is limited. Situations exist where replicate testing to the extent of inserting just two samples from a pool is only tolerated when there is a need to monitor the spread of results about their mean. An example is contained in a reported application of control charts to medical testing by Levey and Jennings. They describe the assessment of a method for measuring urea in blood using control charts based on duplicate samples taken from a pool. Maximizing the life of a pool using minimal sampling levels has to be achieved by formulating control schemes which fully utilize information contained in the test data. Furthermore, when testing is intricate and expensive we only want to interrupt in control situations infrequently. On the other hand we need quick detection when the test level drifts away from a specified target value, or test variability deteriorates. We shall see that the formulation of test methods which meet these objectives require the definition of a criterion called the 'minimal envelope'. Basically this is a yardstick which can be used to compare characteristics of a whole variety of control schemes. Its use has led to identifying schemes which ensure specified average test runs when testing is in control and the speediest detection when it is not.

The test run or run length of a control rule is the number of samples taken in a single use of the rule, between starting it off and reaching an out-of-control decision. Its average value, called the average run length L, is the mean number of control samples which will be tested before the scheme indicates testing is out of control. To illustrate its use in the design of schemes let us introduce the following notation. Let θ denote the parameter being controlled. The average run length (ARL) of a scheme is evidently a function of θ. We will therefore write it as $L(\theta)$. If θ_r is a value of θ which should be rejected, we require $L(\theta_r)$ to be small, whilst if θ_a represents an acceptable value of θ, then $L(\theta_a)$ needs to be large.

1.7 Average run length

Consider the use of ARL in an industrial context where there is freedom to choose any reasonable sample size n and a clinical situation (say) where the choice of n is limited. In the first instance we shall see that a whole variety of schemes can be found with particular values of $L(\theta)$ at θ_a and θ_r. Some of them will, in a given sense, be better than others. If the value of n has to be small we shall find that we may not be able to achieve both of the values $L(\theta_a)$ and $L(\theta_r)$. How can we deal with this situation? One way is to use the largest value of n that we can and determine the smallest value of $L(\theta_r)$ that this permits. We then look for the control scheme which maximizes $L(\theta_a)$.

Choices of scheme

What choices are available between control schemes with common n and $L(\theta_r)$ values? The alternatives are

• the use of different control techniques;

- the identification of parameter values for particular types of scheme which maximize $L(\theta_a)$;
- the use of different functions of the sampled data;
- varying the interval lengths between which samples are taken.

We find that use of these alternatives significantly increases the values of $L(\theta_a)$ which can be achieved for a given value of $L(\theta_r)$.

1.8 Control techniques

What are the different methods of control whose characteristics we can compare? We shall be concerned with four possibilities, namely,

- simple control charts with action lines only;
- Shewhart type charts having warning and action lines;
- cumulative sum charts;
- schemes designed to combine the best features of cumulative sum and Shewhart charts.

Shewhart charts

A typical double-sided Shewhart chart is indicated in Fig 1.2. Samples of size n are taken at regular (equally spaced) intervals. A statistic $S(X)$ is calculated from the sampled data. Values of $S(X)$ are plotted as they become available, in the manner shown in the diagram.

Vertical distances from the horizontal scale represent values of $S(X)$. The property being controlled could be the mean level of a particular clinical test. To simplify our description of the Shewhart chart assume, for the moment that we wish to control the mean value of $S(X)$ to the value θ_a, and that neither an increase or decrease in θ is acceptable. In this case the inner line of the chart is the value θ_a and the lines on either side of it are action or control lines. As long as plotted values of $S(X)$ are randomly distributed about the inner line and remain inside the control lines, testing is judged to be in control. An out-of-control decision is reached as soon as one value of $S(X)$ lies outside them. Values of $L(\theta_a)$ and $L(\theta_r)$ are determined by the distances of the control lines from the central one and the size n of the control samples.

Charts with warning and action lines

Clearly, an out-of-control decision using a Shewhart chart is based on the information contained in just one sample. This is a weakness of the method, particularly when changes in test, or quality levels, are likely to be gradual. Sample values before the one giving the out of control decision may indeed be indicating that a change has

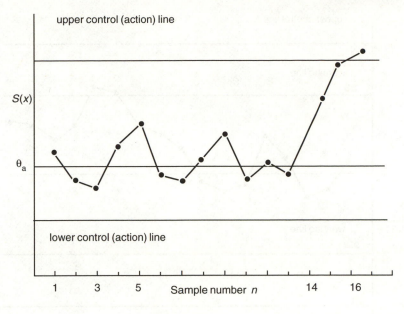

Fig. 1.2   Shewhart chart

occurred. The sample values plotted in Fig 1.3 illustrate such a situation. A run of nine values of $S(X)$ lie above the θ_a line; all of them could be indicating that the expected value of $S(X)$, namely θ_a may have changed. The data indicates this may have occurred at about the time the ninth sample was taken. However, the control rule does not give an out-of-control decision until the eighteenth sample. To some extent we can overcome this weakness of Shewhart charts by introducing additional decision lines into the chart. The new lines are called warning lines and are shown in Fig 1.3. The modified rule could be as follows: take the decision that testing or quality is out of control when a value of $S(X)$ lies outside the upper or lower control lines or if two successive values of $S(X)$ lie between a warning line and action line. Superficially there would appear to be a wide variety of alternative decision procedures of this kind. For example, rather than two successive $S(X)$ values between these lines we could require any three of four successive ones to do so. We can also vary the positions of the lines and their number. The values of $L(\theta_a)$ and $L(\theta_r)$ will evidently change with their positions. For a particular sample size and specified $L(\theta_r)$ we can maximize $L(\theta_a)$ by changing,

- the values of d_1, d_2, d_3, and d_4;
- the rules about points between or outside decision lines;
- the number of decision lines.

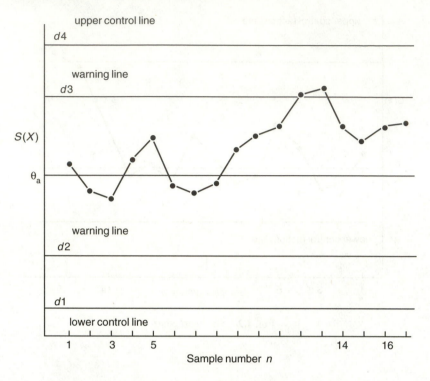

Fig. 1.3   Shewhart chart with two sets of decision lines.

Cumulative sum charts

The introduction of warning lines into Shewhart charts is an attempt to use data more
effectively, in particular for situations where gradual changes in θ are likely to occur.
However, as we shall see, they only partially achieve this objective. We would make
better use of the data if we could devise a technique which accumulates information
about a change in the value of θ immediately after the change, and only if this is in an
out-of-control direction. A relatively new procedure which does so is the cumulative
sum or cusum chart. It is based on selective summations of sampled data. Consider
the sum

$$S_R = \sum_{i=1}^{i=R}(X_i - \theta_a)$$

where X_i is the value of the ith sample. S_R is thus the cumulative sum of the difference
between each control sample value and the 'target value' θ_a. The graph of S_R plotted
against R is called a cusum chart. As the mean value of X_i is θ_a the plotted points
on the chart will fluctuate about the horizontal line representing θ_a. If the mean of
X_i changes the path direction will change. The frequency with which the differences

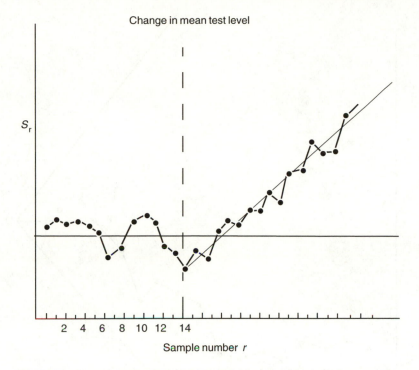

Change in mean test level

S_r

2   4   6   8   10 12   14

Sample number r

Fig. 1.4   Cumulative sum chart.

$(X_i - \theta_a)$ exceed zero increases when the mean of X_i increases from θ_a to a new value of θ. The path of the chart accordingly begins to slope upwards, if the mean test level drops below θ_a it starts to go in a downwards direction. Evidently when testing is on target the mean value of S_R is zero. If, on the other hand, the mean value of X_i changes to θ_r from $R = R_1$ onwards, plotted values of S_R begin to fluctuate about the line $(R - R_1) \cdot (\theta_r - \theta_a)$ indicated in Fig 1.4. Changes in values of θ from plotted data are visually more obvious in Fig 1.4 than from Shewhart type charts. Used carefully this feature of cusum charts can be a valuable tool in the search for causes that lead to changes in the value of θ. Visual inspection indicates both changes in level and the approximate point in time at which they occurred. The value of this particular aspect of the cumulative sum charts cannot be over-emphasized, both in the operation of complex industrial processes and in clinical testing. Suppose the procedure being controlled is clinical and is being used for disease diagnosis; the chart can be utilized to flag patient samples which should be re-assessed. The cusum technique can therefore fulfil more than one role.

One way to turn Fig. 1.4 into a control chart, is with the use of a V-mask to signal changes in θ. The super-imposition of this device on a cusum chart is shown in Fig 1.5. It can be seen that the line OP bisects the angle 2α between the limbs of the 'V'.

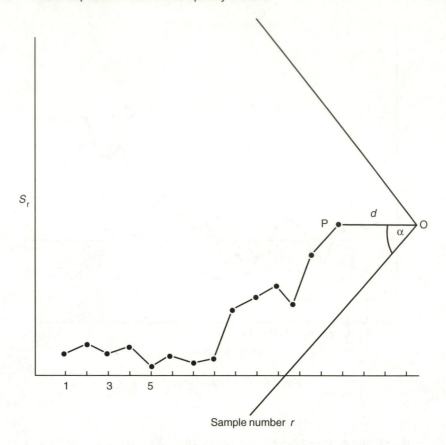

S_r

P

d

α

O

1   3   5

Sample number r

Fig. 1.5   V-mask.

The point P is distance d from O, where d is in the units used for the vertical scale to plot the values of S_R. The mask is used by placing it on the chart with OP horizontal and P coinciding with the last plotted point. If the path of plotted points crosses either limb of the V conclude that θ has changed. Different values of $L(\theta_a)$ and $L(\theta_r)$ are obtained by varying d and α. The cusum method of control requires less test data (smaller sample sizes) to achieve particular values of $L(\theta_a)$ and $L(\theta_r)$ than Shewhart charts. This is because they use more of the information contained in sampled data.

We do not need to plot all of the sample values obtained, as in Figs 1.4 or 1.5, we can operate a chart similar to a Shewhart chart by defining a parameter k called the reference value. This parameter takes values between θ_a and θ_r. Let us consider a situation where we are only interested in detecting increases in θ, so that $\theta_r > \theta_a$. The procedure is to start summing results only when a sampled value X_i is greater than $\theta_a + k$. Suppose the first occasion that this happens is when $i = R_1$. Values of $X_i - \theta_a - k$ are accumulated from this point onwards until either the sum returns to zero or less, or it exceeds a given value h called the decision interval, if we write,

$$S_R(t) = \sum_{i=R_1}^{i=R_1+t} (X_i - \theta_a - k)$$

we calculate $S_R(t)$ until it becomes zero or less or greater than h. If the first happens cease cumulating until the next value of $X_i > \theta_a + k$ occurs at which point the next series of values of $S_R(t)$ commences. This process is continued, cumulation being 'triggered' by the condition $X_i > \theta_a + k$ until the first occasion that $S_R(t) > h$. When this happens conclude testing is out of control. Obviously a similar process is used to detect decreases in θ. To detect both positive and negative movements in θ a second cumulative sum $S_R(T)$ is computed namely,

$$S_R(T) = \sum_{i=R_2}^{i=R_2+T} (X_i - \theta_a + k)$$

where R_2 is the first value of R that $X_i < \theta_a - k$. The calculation of a particular $S_R(T)$ continues until it becomes >0 or $< -h$. If the first happens then, as with the first set of cumulations, the re-calculation of $S_R(T)$ commences at the next value of i for which $X_i < \theta_a - k$. The process of repeated cumulation continues until $S_r(T) < -h$. To test for positive and negative deviations from θ_a, values of $S_R(t)$ and $S_R(T)$ are obtained concurrently. Testing is judged out of control when either $S_R(t) > h$ or, $S_R(T) < -h$. We shall see that this double decision procedure is equivalent to a V-mask.

Combined schemes

A Shewhart chart with action lines only, is known to be insensitive to small changes in θ. Cumulative sum charts were devised to overcome this weakness. Single cusum schemes, for example, can be formulated which are sensitive to changes in θ up to $2\sigma(X)$ away from θ_a, where $\sigma(X)$ is the standard deviation of the statistic X being cumulated. The properties of such a scheme, however, may not be as good as a Shewart chart for changes in θ greater than $2\sigma(X)$. Accordingly the superiority of a cusum scheme for small changes in θ and that of a Shewhart chart for larger increases in θ led to the search for schemes which combine the best features of both. One obvious way of doing so is to consider the simultanious use of a single cusum scheme and a chart with warning and action lines. Considerable attention has been focused on such combinations in the literature concerned with clinical testing.

With regard to the search for the most sensitive schemes it is important to recognize the following point, namely, the observations we have just made do not imply that we cannot design a cusum which is superior to Shewhart type schemes. We can design a cusum scheme with superior properties to a Shewhart scheme for θ greater than $2\sigma(X)$, what we cannot obtain, is a single scheme with better properties both in the region close to the control value of θ, and changes in level of $2\sigma(X)$ or more. In view of this remark, an alternative to the mixture of two different techniques is the simultaneous operation of two appropriate cusum schemes. Various joint schemes have been considered in the literature. Westgard *et al.* (1981) have examined some which combine cusum and Shewhart type charts. Bissell has investigated a procedure which consists of the superimposition of some five different cusum schemes. Neither of these comparisons, however, was based on the concept of the minimal envelope mentioned earlier. When this criterion, which represents the best testing scheme which can be designed, is used we shall see that for all practical purposes, the simultaneous operation of two cusum rules gives control schemes very close to the best that can be achieved.

From the techniques so far described and the parameter determinations required in their design, there is clearly a wide variety of alternatives which need to be assessed for the identification of optimal methods of control. In this respect additional possibilities are available with the choice of which control statistics could be used.

1.9 Control statistics

To illustrate this last observation let us take the control of test error to show that there is indeed an element of selection with regard to the way sampled data can be summarized to obtain control schemes with 'best' features. Suppose the variability of a test procedure is to be monitored by schemes designed to control its standard deviation σ. Denote in- and out-of-control values of σ by σ_a and σ_r respectively. Suppose that the size of each control sample is n and that the jth member of the ith control sample is X_{ij} ($j = 1, 2, 3, \ldots, n$). Suppose also, we want schemes with fixed $L(\sigma_r)$ and need to maximize $L(\sigma_a)$. A number of sample statistics related to σ can be used. Without some investigation it is not obvious which will give a better or best scheme. What sample statistics could we use? One possibility is the sample range,

$$S_1(X) = X_{i1} - X_{is}$$

where X_{i1} is the largest value of X_{ij} and X_{is} is its smallest value. On the other hand we could use,

$$S_2(X) = \left[\sum_{j=1}^{j=n} (X_{ij} - \bar{X}_j)^2 / (n - 1)\sigma \right]^{1/2}$$

where

$$\bar{X}_j = \sum_{j=1}^{j=n} X_{ij} / n.$$

Yet another statistic we could use is $S_2(X)$ squared. We shall look at these alternatives later.

Situations arise in clinical testing, for example, where the need to detect out of control testing overrides the need to maximize the life of a control pool. In these circumstances we want schemes with minimum $L(\sigma_r)$. We shall therefore need to search for two kinds of scheme, those with maximum run lengths at AQL for given run lengths at RQL and those with minimum run lengths at RQL for given values at AQL.

1.10 Sampling interval

An important aspect of control testing is the average number of batches of clinical material likely to be processed after testing has gone out of control. Suppose the number of batches tested between successive control checks is b and θ is the parameter being monitored. If the average run length at RQL is $L(\theta_r)$, the average number of batches processed when testing is out of control is $b \cdot L(\theta_r)$. The corresponding number of batches tested when testing is in control is $b \cdot L(\theta_a)$. Can additional maximization of $L(\theta_a)$ and accordingly $b \cdot L(\theta_a)$ be achieved for the same life of the pool by increasing the number n of control samples tested, proportionately increasing b and decreasing $L(\theta_r)$? If the sampling interval is increased to $3b/2$ and $L(\theta_r)$ is decreased to $2L(\theta_r)/3$, the number of batches tested on average whilst testing is out of control is still $b \cdot L(\theta_r)$. Suppose the number of control samples available in the pool at the beginning of its life is N. Its life l in terms of batches which can be tested is $l = Nb/n$. Obviously l does not change if b and n are adjusted by the same proportion, thus we can increase n by 50 per cent and increase b to $3b/2$ without affecting l. We shall examine whether for a specified value of $b \cdot L(\theta_r)$ we can find schemes with maximum values of $b \cdot L(\theta_a)$ by increasing n and the sampling interval b. In practice it is, of course, necessary to compromise between increasing $L(\theta_a)$ and the consequences of decreasing the sampling frequency.

Identifying control rules which are optimal in the above senses is a problem of constrained maximization or constrained minimization. Thus, for the former, for a specified $L(\theta_r)$, we need a search procedure to determine the parameter values of the scheme with the highest achievable value of $L(\theta_a)$. For a cusum procedure this means determining appropriate values of h and k, for Shewhart type control rules we want to know where to place warning and action lines. Computerization of these processes requires subroutines with which to calculate the values of the average run lengths $L(\theta_a)$ and $L(\theta_r)$. We shall see that this is a comparatively straightforward matter for Shewhart schemes but rather more complex for cumulative sum schemes. The determination of the run lengths of some schemes which combine both methods of control is very complicated from a theoretical point of view. They have accordingly been obtained using methods of simulation. When doing so it is necessary to take great care in the generation of long strings of 'pseudo' random numbers!

1.11 Randomness

When a sample of items is taken from a production unit, the information obtained from measurements made upon them is used to assess the quality of articles being produced by the plant. If we are to use the methods of statistics to do so we must ensure that each item is drawn randomly from the population of items being produced. This means that each item is selected non-systematically; that is, in a haphazard manner. The term 'simple random sample' is used to describe a particular kind of sample commonly used in statistical investigations. In such a sample, the drawing of each successive item is independent of what has happened before, and further more, every item in the population has the same chance of selection. Although items are drawn at random, in practice they are sampled from a production unit without replacement. The probability that a particular item is selected therefore depends upon those which have been sampled before it. Clearly, such a sample is not a simple random sample. Statistical theory based on simple random sampling is much more straightforward than that of non-replacement sampling. The difficulties of the latter can, of course, be overcome by assessing each item and then returning it to the population from which it came before selecting the next item. Such a procedure is neither sensible nor practical in a quality control context. If n the number of items in the sample is small relative to N the population size, we can take the theory of simple random sampling to be an adequate approximation to that of random non-replacement sampling. Thus suppose we wish to control the proportion of defective items contained in batches of 5000 items sold to customers. It is proposed to do so by testing 50 items from a batch. Suppose further, that batches with 5 per cent or less defectives are acceptable whilst those with 15 per cent are not. If p the proportion of items in the batch is 0.05, theory we shall shortly describe shows that we are unlikely to obtain more than five or six defective items in a sample. If, on the other hand, p is 0.15 we are unlikely to find more than 15 defective ones. Thus, if $p = p_a = 0.05$ the proportion of defective items remaining in the sampled batch is not likely to be less than 0.049 whilst if $p = p_r = 0.15$ then p will only rarely be less than 0.148 after sampling. In either case the changes in p attributable to non-replacement sampling are very small. In such circumstances it is not surprising to find that the difference between decision rules based on non-replacement sampling, and those based on simple random sampling are of no practical importance. The theory which follows in this monograph is based on this assumption. For convenience we shall take the term random sampling to be synonymous with simple random sampling.

Taking a sample

A straightforward way of drawing a random sample from a population is with the use of tables, which have been constructed so that each number of them has the same chance of selection. A typical set of random numbers between 0 and 100 is shown in Appendix Table I. To illustrate its use suppose we have a production unit consisting of 100 machines and, for control purposes, we want a random sample of 10 of them taken at regular intervals. Suppose also that we want to ensure that all of the machines get tested in due course. To do this we merely number the machines from 1 to 100 and, starting at a random point in the Table take 10 numbers across it and in any set of 10 disregard any number that has already occurred until all numbers between 0 and 100 have been used. Thus for the first sample we could start at the third row and second column and use the numbers 20, 33, 64, ..., 18

It is sometimes necessary to generate a very large number of random numbers particularly for simulation purposes like those described at the end of this monograph. As already remarked, the method of generating these has to be carefully chosen. A procedure for the computerized generation of numbers which can represent randomness is called a mixed congruential multiplier. Although this may seem a rather frightening term to the non-mathematician it is, in fact, a very simple procedure. Suppose the nth number in the sequence is x_n then

$$x_{n+1} = (\lambda x_n + \mu) \bmod P,$$

is the notation used for the residual left after $(\lambda x_n + \mu)$ has been devided by P. To illustrate the method, suppose $\lambda = 5$, $\mu = 7$, and $P = 2^4 = 16$ then if x_1 is 1 the series of numbers generated by the above equation is 1, 12, 3, 6, 5, 0, 7, 10, 9, 4, 11, 14, 13, 8, 15, 2, 1. The method appears to give a random arrangement of all numbers between 0 and 15 with each number occurring only once. Notice that after $n = 16$ the sequence just given will repeat itself. Sequences of numbers with such random appearance can be generated and used to design both small and large sampling plans provided the values of λ, μ, and P satisfy certain conditions. Evidently if we take the generator

$$x_{n+1} = (\lambda x_n + \mu) \bmod P^r \tag{1.1}$$

can generate a large series of numbers when $P = 2$ and r is large, say 35. Strictly speaking, this and similar methods do not produce truly random numbers since the sequence is completely specified once x_1, λ, μ, P and r are chosen. The method produces numbers which in many ways behave like random numbers they are accordingly called 'pseudo random numbers'. Computer programs based on methods like the one we have just described have been devised for use in practice. They have been subjected to batteries of statistical tests to ensure randomness before being released. Such sets of numbers have been used in the simulations described in Chapter 7.

1.12 Statistical method

For readers unfamiliar with the methods of statistics it may be useful here to draw attention to some salient features of its methods, the circumstances when they are needed, and the consequences of using them. It is particularly apt to do so in the field of quality control.

It is a surprise to many people to be told that statistical method, viewed as a mathematical discipline, plays a paramount role in modern research and technology. In what sense is this so? Its methods are essential when

- there is inherent uncertainty with regard to the occurrence or non-occurrence of the phenomena being investigated;
- measurements taken contain unavoidable components of error;
- the number of observations which can be made leading to a particular decision on a given population is limited.

To take an example, there are many situations in medical research where progress can only be made from experiments involving small numbers of observations. The phenomena being investigated may or may not occur in a single experimental unit. There will therefore be differences in the outcomes of different groups of experiments or sets of observations because of this element of chance. Experimentation is concerned with altering the rate at which chance events occur. Frequently in medicine the object is to increase the chance or probability of cure in the treatment of a disease.

Another type of variation between experimental results arises because it is not possible to assign accurate measurements to observed events. Measurement involves an inescapable element of error called a random component of variability. There are many circumstances where even with the use of highly developed instrumentation it is only possible to obtain approximate values of the phenomenon being examined.

Statistical theory is concerned with the conclusion which can be drawn when either or both of these sources of variability are present, and where the number of observations which can be made has to be small. This aspect of statistics is called small sample theory. We shall see that probabilities can be assigned to the various outcomes which may arise. These can be used to assess the consequences of accepting or rejecting different theories.

1.13 Right and wrong decisions

Consider the control of the proportion of defective items produced by a manufacturing unit, where the only way we can assess the proportion of defective items being produced is by taking a small sample. Let us return to the example where N the size of a batch is 5000 and $p \leq p_a = 0.05$ represents acceptable quality and $p \geq p_r = 0.15$ rejectable quality. A random sample of n items is taken without replacement from a batch, and we use the rule, if the number of defective items in the sample is $<k$ conclude the batch is one which contains 5 per cent or less defectives. If the number is $\geq k$ conclude it has 15 per cent or more defectives. If $n = 50$ and $k = 5$ we find, from an expression we shall derive in Chapter 2, that in the long run 10 per cent of control samples taken from the process will have five or more defective items when $p = 0.05$. Prolonged use of the rule would therefore result in the rejection of acceptable batches 10 per cent of the time. On the other hand, it would result in the rejection of 89 per

cent of batches with $p = 0.15$. Applying this particular rule the chance of accepting such a batch would be 1 in 11. Thus for both values of p correct decisions would be taken about 90 per cent of the time and incorrect ones on 10 per cent of occasions. As far as the layman is concerned it is important to emphasize that the use of statistical decision rules inevitably incurs the risk of reaching wrong conclusions. We cannot eliminate this possibility; we can, however, design schemes to keep the frequency of reaching wrong decisions to acceptably low levels. We shall see that this is so when we consider the present example in Chapter 2.

2
Small samples: Decisions and consequences

2.1 Small samples

In this chapter we shall indicate the possibilities of small sample theory in more detail than Chapter 1. We shall define and discuss concepts needed to make decisions on the information in them, and the consequences of their use in quality control. Small sample theory can be summarized as rational procedures for drawing inferences about the nature of a population based on limited data sampled from it. Broadly speaking, its methods relate to two kinds of situation. The first is concerned with events which do not always occur when a sample of just one item is taken from the population; that is, they have a 'probability' of occurrence. The second relates to circumstances where measurements made of a specific property of a sampled item, or the property itself, is subject to an unavoidable component of variability.

To discuss the first, let us centre attention on situations where there are only two possibilities with regard to the outcome of a sample of one item, namely, either a defined event occurs or it does not. An example is the single toss of a coin; if it has been properly minted it will fall with head or tail uppermost and we cannot predict which of these two events will occur. Many phenomena in the real world are of this kind. Take, for example, the emission of particles from a radioactive source. Observations over short fixed time intervals will reveal the emission of a particle during some intervals and not in others. The emission of a particle appears to be random in time. In the treatment of some diseases, we find that some patients will respond to new drugs whilst others will not. To study an important category of diseases it is necessary to measure particular hormone concentrations in a given system. Direct measurement of the hormone is not possible, its presence can be detected only if it becomes attached to a specific chemical reagent when it is introduced into a test sample. Sometimes this happens in a small unit of volume of the reagent and sometimes it does not. The development of quantitative methods, based on the information contained in small samples, has been a necessary prerequisite to the advancements we have seen in science, technology, medicine, and modern manufacturing methods.

In reality there are, of course, situations where the outcome of a unit sample has more than just two possibilities. In the context of industrial quality control an article may be defective for a number of reasons. For example, a bobbin of the synthetic yarn we have discussed in Chapter 1, could be defective because it would produce ringy hosiery, a stripy fabric, or the yarn on it may have a number of dyeing deficiencies, or the finish on the yarn may be faulty, producing an unacceptably high static charge during a knitting process. Methods for dealing with such circumstances are obviously more complicated than those which relate to whether or not an item is or is not acceptable due to just one cause. Appropriate control procedures can be devised once methods for this simple situation have been formulated. We shall not, however, deal with them in this monograph.

The second kind of uncertainty is easier to describe if, at first, we confine attention to measurement errors. Let us consider a situation where it is not possible to accurately measure the property of the item we want to control. The measurements we make have an inbuilt error in them. If x_t represents its true value and x_s is its sampled value, then x_t is unknown. The value obtained for the unit sample is $x_t + \varepsilon$, where ε is a measurement error whose value we do not know. What is known about ε is that it takes both positive and negative random values which differ from one value of x_s to the next. Thus, two values of x_s with a common x_t would be $x_t + \varepsilon_1$ and $x_t + \varepsilon_2$, ε_1 and ε_2 being two separate values of ε.

To illustrate some basic principles which we use to deal with such sample values x_s, let us make some assumptions about ε. Suppose it takes values between $-\alpha$ and $+\alpha$, and that it takes any value in this range equally often. If a very large number of values of x_s are obtained, all with a common value x_t, then if the proportion of times $p(x_s)$ that x_s takes a particular value of x between $x_t - \alpha$ and $x_t + \alpha$ is recorded, then the diagram of $p(x)$ plotted against x will resemble Fig. 2.1.

If we only have one sampled value $x_s = x$, what can we say about the unknown value of x_t? If the value of α is known, we can say with certainty that x_t lies between $x - \alpha$ and $x + \alpha$. Suppose the value of α is much too large for this last statement to be of any practical use with regard to obtaining an estimate of the unknown value of x_t. Is it possible to obtain information from a relatively small number of values of x_s, to draw conclusions about the unknown value of x_t which can be used in practice? Let us examine this question by supposing that we have n randomly sampled values of x_s. Denote the particular value of x_s obtained for the ith sample by x_i, so that if ε_i is its value of ε, $x_i = x_t + \varepsilon_i$ and

$$\sum_{i=1}^{i=n} x_i = nx_t + \sum_{i=1}^{i=n} \varepsilon_i. \tag{2.1}$$

With regard to the sum of the random errors ε on the right-hand side of this equation, there will usually be a fairly equal mixture of positive and negative values of ε in the n-fold sum of x_i. Thus the sum of the ε values will cluster about the value 0, and will lie well away from the extremes $-n\alpha$ and $n\alpha$. This argument leads us to expect that the mean value of the sampled values x_i, will in general be close to x_t, if n is

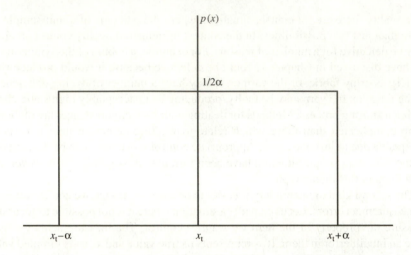

Fig. 2.1 Proportion of $p(x)$ of values of x when ε is evenly distributed between $x_t - \alpha$ and $x_t + \alpha$.

sufficiently large. If we write

$$\bar{x}_s = \sum_{i=1}^{i=n} x_i/n = x_t + \sum_{i=1}^{i=n} \varepsilon_i/n = x_t + \bar{\varepsilon}, \qquad (2.2)$$

it seems reasonable to expect that on average the difference between the sample mean and x_t will become smaller as n becomes larger. If this speculation is correct, an important question now arises in the context of quality control, namely, what sample sizes should we take for the mean of the x_s values to be close enough to x_t to use it as an estimator of the unknown value of x_t?

An Illustration

Either by using tables of random numbers, or, more easily, a method of generating pseudo random numbers, such as that described in Chapter 1, we can get a feel for the way in which mean sample values centre around x_t when n is relatively small. In doing so, we shall see that increased clustering of sample means about x_t does indeed occur as n increases. In addition, a significant feature of the distribution of the mean value of ε about the point $x = x_t$ also becomes apparent. We shall see that this aspect of its behaviour is an example of an important theorem in statistical theory. To illustrate all of these points, groups of n experimental values of x_i were obtained by adding computer-generated random values of ε to the value $x_t = 10$; the range of values of ε was fixed as -5 to $+5$ so that $\alpha = 5$. Three separate lots of experimental results were obtained. Each experiment consisted of 200 sets of values of n, with n being 10, 20, and 30. The mean values of x_s were calculated for each individual

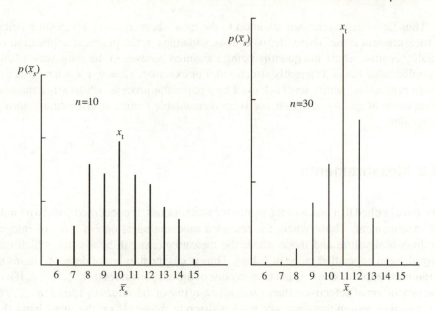

Fig. 2.2 Values of $p(\bar{x}_s)$ plotted against $\bar{x}_s$.

sample, so that 200 means were obtained for each sample size. The percentage of them which differed from x_t by 1.5 or less was 50 per cent for $n = 10$, 70 per cent for $n = 20$ and 80 per cent for $n = 30$. The proportion of values obtained in the ranges 5.5 to 6.49, 6.5 to 7.49 up to 14.5 to 15.49 are shown in Fig. 2.2. The 'significant feature' to which we have just referred is that each distibution has an inverted bell shape, which becomes more symmetrical about the point $x_t = 10$, as n increases in value. It is also clear from Fig. 2.2 that as n increases the width of the 'bell' becomes narrower and more peaked. This obviously indicates that as n increases so sampled averages become better estimates of x_t. We would, of course, intuitively expect this to be so. It seems, however, from our experiment, that n does not necessarily have to be very large for x to be close to x_t in a large percentage of samples.

The bell shape about $x_t = 10$ indicates the manner in which mean values of ε vary about zero. If we repeated this type of experiment for a wide variety of different distribution shapes for ε, we would also find that the distribution mean values of ε about the mean point x_t is bell-shaped, and narrows with increasing n. As we shall see, this feature is central to the mathematical basis of statistical theory. It is an indication of a theorem known as the central limit theorem, which we shall consider in Chapter 4. With its use we can formulate the statistical distribution of mean values of quantities like x_s and ε. Subject to just one condition we can do this irrespective of the shapes of the distributions of individual values of x_i and ε_i. This is an exceedingly significant, if astonishing result! For it means that whatever the distribution of x_i likely to occur in practice, we do not need to know what it is in order to determine the distribution of mean values of it.

Thus far we have confined attention to the case where $x_s = x_t + \varepsilon$, with ε being a measurement error. There are, of course, situations in the practical application of quality control, where the quantity being measured behaves in the same way as this sampled value but is differently structured. For example, $x_s = \theta + \varepsilon$ where θ is the mean production quality level achieved by a particular process, whilst ε is a random component of quality variation and is an unavoidable feature of the manufacturing procedure

2.2 Measurements

We have implied that much of the theory we shall discuss, is concerned with two kinds of measurement. Those where the recorded outcome takes one of a set of integer or fractional values and those where the measurements can be a value which lies anywhere in a specified range of values. Thus if x is the number of defective items in a random sample of size n, it can take any one of the integer values $0, 1, 2, \ldots, n$. If it is the proportion of defectives, then x will take any one of the values $0, 1/n, 2/n, \ldots, 1$. In statistical terminology we say x takes discrete values. If on the other hand the quantity being measured is the sum of a true value such as x_t and a measurement error ε, which can take any value at random, we say it is continuous.

2.3 Events

We now need to define different kinds of event and the manner in which they relate to one another. We shall want the following terms:

- mutually exclusive events;
- exhaustive classification;
- favourable event.

Mutually exclusive event

Two events are mutually exclusive if the occurrence of one excludes the other. For instance, when sampling the output of a manufacturing plant, the occurrence in a unit sample of a non-defective item excludes that of a defective one.

Exhaustive classification

The classification of possible outcomes is exhaustive if it covers all of the possibilities which can arise. Classifying articles as defective or non-defective is obviously exhaustive. If the score attributed to the single toss of a six-sided dice is the number of pips on the uppermost face, then the values 1, 2, 3, 4, 5, and 6 constitute an exhaustive classification.

Favourable event

The number of outcomes favourable to an event E is the total number which can result in E. If a six-sided dice is tossed twice, the 36 pairs $(1, 1), (2, 1), (3, 1), \ldots, (6, 6)$ constitute an exclusive set of paired values. If we add the two scores together we find that four of them sum to 5. The number of events favourable to a score of 5 is therefore 4. Just two outcomes, namely $(6, 5)$ and $(5, 6)$, lead to a score of 11, so that two events are favourable to this result.

2.4 Probability

Suppose we have n mutually exclusive and exhaustive outcomes, all of which occur at the same rate. If r of these are favourable to E, then the probability $p(E)$ that E will occur is (r/n). Thus, for two tosses of the dice just considered, there are 36 mutually exclusive pairs of possible scores and four of them are favourable to a score of 9, so that the probability of obtaining a score of 9 in two tosses is $1/9$. If we have a perfectly made coin, there are only two exclusive possibilities, one of which is a head. The probability of a head in a single toss of the coin is evidently $1/2$. If there are 50 defective items which are randomly distributed in a batch of 5000 manufactured items, the probability of sampling a defective article is $1/100$.

Total probability

With these definitions and two relationships between the probabilities of separate events, we can determine the probabilities of combinations of them. To establish the first, let us consider a situation where any one of n mutually exclusive events is equally likely. If r_1 are favourable to event E_1 and r_2 to a different event E_2, then when E_1 and E_2 are mutually exclusive, the number of events favourable to either E_1 or E_2 is $(r_1 + r_2)$ so that

$$p(\text{either } E_1 \text{ or } E_2) = (r_1 + r_2)/n = p(E_1) + p(E_2), \qquad (2.3)$$

and evidently

$$p(E_1 \text{ or } E_2 \text{ or } \ldots E_j) = \sum_{i=1}^{i=j} p(E_i) \quad (j < n)$$

in particular

$$p(E_1 \text{ or } E_2 \text{ or } \ldots E_n) = \sum_{i=1}^{i=n} p(E_i) = \sum_{i=1}^{i=n} r_i/n = 1.$$

For the twofold toss of a dice the number of events leading to a score of 7 is 6, and for a score of 8 is 5. The two scores are mutually exclusive, so that the number of events favourable to either is 11 and we have

$$p(7 \text{ or } 8) = 11/36 = p(7) + p(8).$$

The second relation we require permits the determination of the joint probability of several events. To establish it, we need to distinguish between dependent and independent events. A simple way of doing so is to consider the selection of cards from a pack of 52 playing cards. Suppose two cards are taken from a pack and that the second is taken without the first being replaced. What are the probabilities of obtaining cards of red or black suit? If E_1 is a red suit for the first card taken then $p(E_1) = 1/2$. If E_2 is the occurrence of a red suit for the second card, there are 51 cards remaining in the pack with 25 or 26 red ones depending upon the nature of the first card drawn. Clearly, the probability of E_2 is conditional on whether E_1 occurred or not. If we use $p(A \mid B)$ for the probability that A occurs on condition B has already occurred, then

$$p(E_2 \mid E_1) = 25/51.$$

If we use E_1' to denote the non-occurrence of E_1 then

$$p(E_2 \mid E_1') = 26/51.$$

If the first card is replaced in the pack at random before the second is taken, the probability $p(E_2)$ does not depend on whether or not E_1 occurred. E_2 is independent of E_1. The dependence or independence of E_1 and E_2, in our example, arises from the method of sampling. There are many other ways in which dependent events occur in practice. The example given suffices to demonstrate the need to distinguish between these two different kinds of event, if we are to obtain expressions for their joint probability in terms of their separate probabilities

Joint probability

Consider the case where of n mutually exclusive outcomes, r_1 are favourable to E_1, r_2 to E_2 and r_3 to both E_1 and E_2, so that

$$p(E_1 \text{ and } E_2) = r_3/n, \quad p(E_1) = r_1/n, \quad p(E_2) = r_2/n,$$

and

$$p(E_2 \mid E_1) = r_3/r_1 \quad \text{and} \quad p(E_1 \mid E_2) = r_3/r_2.$$

Since we can write $r_3/n = (r_1/n) \cdot (r_3/r_1) = (r_2/n) \cdot (r_3/r_2)$ it follows that

$$p(E_1 \text{ and } E_2) = p(E_1) \cdot p(E_2 \mid E_1) = p(E_2) \cdot p(E_1 \mid E_2). \tag{2.4}$$

We therefore see that the probability of two dependent E_1 and E_2 is the product of the probability of E_1 and the conditional probability of E_2, E_1 having occurred, or equivalently, the probability of E_2 times the probability of E_1 on E_2

Statistical independence

If two events are independent of each other, with r_1 of n mutually exclusive and favourable to E_1, and r_2 favourable to E_2, then, the number of events favourable to both E_1 and E_2 in a twofold trial is $r_1 r_2$. Since the total number of events in such a trial is n^2 we have

$$p(E_1 \text{ and } E_2) = (r_1 \cdot r_2)/n^2 = p(E_1) \cdot p(E_2).$$

The probability that two independent events occur is the product of their separate probabilities. We are led to the following definition: two events are statistically independent if their joint probability is equal to the product of their separate probabilities. Evidently the joint probability of i independent events is the product of their separate probabilities, so that

$$p(E_1, E_2, \ldots, E_i) = \prod_{j=1}^{j=i} p(E_j).$$

Total probability of dependent events

We have established that for two mutually exclusive events the probability of either is the sum of their separate probabilities. What is the corresponding result when the occurrence of one does not exclude the other; that is, two of a set of events, namely, E_1 and E_2, are not independent? As above, let us use E' to denote the event 'not E'. The events E_1 and E_2, and, E_1 and E'_2 are mutually exclusive, so that

$$p(E_1) = p(E_1 \text{ and } E_2) + p(E_1 \text{ and } E'_2) \tag{2.5}$$

and

$$p(E_2) = p(E_2 \text{ and } E_1) + p(E_1 \text{ and } E'_1). \tag{2.6}$$

To establish the probability of E_1 or E_2 we need these two results and the combinations

- C_1 which is E_1 and E_2;
- C_2 which is E_1 and E'_2;
- C_3 which is E_2 and E'_1.

Evidently C_1 and C_2 are mutually exclusive and the two together exclude C_3, so that

$$p(E_1 \text{ or } E_2) = p(C_1 \text{ or } C_2) + p(C_3) = p(C_1) + p(C_2) + p(C_3).$$

Using this result and eqns (2.5) and (2.6) gives

$$p(E_1 \text{ or } E_2) = p(E_1) + p(E_2) - p(E_1 \text{ and } E_2). \tag{2.7}$$

This important result gives a general expression for the probability of the occurrence of either of two events in terms of their separate and joint probabilities irrespective of whether they are independent or not. Thus, suppose a perfect dice is thrown twice. What is the probability of a total score of more than 8, when the result of the first toss exceeded 4? If E_1 is scoring more than 4 for the first toss, and E_2 is a score >8 for the two tosses, then E_1 and E_2 are not mutually exclusive. If we list all of the score combinations, we find the number of ways E_1 or E_2 can arise is 15, so that $p(E_1 \text{ or } E_2) = 5/12$. Alternatively, $p(E_1)$ is $1/3$, $p(E_2) = 5/18$ and $p(E_1 \text{ and } E_2) = 7/36$, substitution in (2.7) gives the same result. Equation (2.3) can be obtained from (2.7), for if E_1 and E_2 are mutually exclusive $p(E_1 \text{ and } E_2)$ is obviously zero. If E_1 and E_2 are not mutually exclusive but are independent then,

$$p(E_1 \text{ or } E_2) = p(E_1) + p(E_2) - p(E_1) \cdot p(E_2). \tag{2.8}$$

Example

Two components c_1 and c_2 are required for the production of a particular article, these are supplied to the manufacturer by four different suppliers. Twenty-five per cent of each component held in his warehouse come from each of these separate sources. If one of each component is taken at random from the warehouse, what is the probability that at least one of the pair came from the fourth supplier?

Let us list the combinations of sources S_1, S_2, S_3, and S_4 from which the components c_1 and c_2 can come; for example both c_1 and c_2 could come from S_1. If we denote this by $S_1 S_1$, the pairs containing at least one S_4 would be 7, so that p(at least one component from S_4) $= 7/16$, whilst $p(c_1 \text{ from } S_4) = 1/4$ and $p(c_2 \text{ from } S_4) = 1/4$, so that using eqn (2.8) gives the same result.

Finally, for a pair of dependent events E_1 and E_2, we can express eqn (2.7) in a similar form to (2.8) since

$$p(E_1 \text{ and } E_2) = p(E_1) \cdot p(E_2 \mid E_1),$$

so that

$$p(E_1 \text{ or } E_2) = p(E_1) + p(E_2) - p(E_1) \cdot p(E_2 \mid E_1)$$
$$= p(E_1) + p(E_2) - p(E_2) \cdot p(E_1 \mid E_2). \tag{2.9}$$

2.5 Definitions

In the development of statistical theory we need to use the term variable in two different ways. The first is for the manipulation of functional relationships of the kind

$$y = (2\pi)^{1/2} \exp(-x^2/2).$$

Here y and x are variables which define the above expression, y takes values in the range 0 to one, and x can take any value between $-\infty$ and $+\infty$. To develop a 'calculus' of probability based on eqns (2.3) to (2.9), we need a notation which distinguishes between a particular value of x in this range, and any of the values that x can take. The term random variable or variate is used when referring to generalized values of such a variable. To take an example, suppose we assign the value 1 to an article sampled from a batch of manufactured items if it is defective, and zero if it is not. If the proportion of defectives in the batch is p, this scoring system defines a variate which takes the values 0 and 1, with probabilities $(1 - p)$ and p. Clearly, a random variable measures the outcomes of chance events. Examples are crop yield in agricultural field trials, the wear of a woven fabric, the life of a car tyre or electronic components, the number of patients responding to treatment in a clinical trial. Functions of random variables are themselves random variables, thus the average of n variate values is a variate. It is customary to use a capital letter to denote a variate, X say, and to use the corresponding small letter x for a particular value of X. Thus we represent the variate defined as the mean of n variates by

$$\bar{X} = \sum_{i=1}^{i=n} X_i/n,$$

and the mean of a particular set of observations by

$$\bar{x} = \sum_{i=1}^{i=n} x_i/n.$$

Two pieces of information are needed to define a variate X. They are the range of values it can take, and the probability $p(x)$ that X has a particular value x. Thus for a six-sided dice, the score X obtained from a single toss takes the values 1 to 6 and the probability attached to a particular value of x in this range is $1/6$, whilst for values outside this range $p(x) = 0$. With these definitions and the equations of Section 2.4 formulating the probability of x defectives in a random sample of n items is straightforward. From them it follows that the probability of obtaining x defectives and $(n - x)$ non-defectives in a particular order is $p^x \cdot q^{(n-x)}$ where $q = (1 - p)$. If we now take account of the number of alternative orders which give rise to this number of defectives in a sample of n, then the total probability of x defectives $p(x)$ is

$$p(x) = {}^nC_x \cdot p^x \cdot q^{(n-x)}. \tag{2.10}$$

Frequency function

It is convenient to adopt the convention that all variates take all values between $-\infty$ and $+\infty$. We do so by attaching a value $f(x)$ to each value of x in the range $-\infty$ to $+\infty$, where $f(x)$ takes positive or zero values only. Thus to define a variate all we need is the values of $f(x)$ for all x. Hence, for the score obtained by the single toss of a perfect six-sided dice the variate is

$$f(x) = p(x) = 1/6 \quad \text{for } x = 1, 2, 3, 4, 5, \text{ or } 6,$$
$$f(x) = 0 \quad \text{for all other } x,$$

and for a variate with just two values 0 and 1 with probabilities q and p,

$$f(x) = p(0) = q \quad \text{for } x = 0, \qquad f(x) = p(1) = p \quad \text{for } x = 1$$

and

$$f(x) = 0 \quad \text{for all } x \neq 0 \text{ or } 1.$$

We shall abbreviate this notation further and write, for example,

$$f(x) = 1/6 \quad x = 1, 2, 3, 4, 5, \text{ and } 6$$
$$= 0 \qquad \text{otherwise},$$

so that the variate defined by (2.10) is

$$f(x) = p(x) = {}^{n}C_{r} p^{x} q^{(n-x)} \quad \text{for } x = 1, 2, 3, \ldots, n$$
$$= 0 \qquad\qquad\qquad \text{otherwise.}$$

Use of $f(x)$ defined in this way simplifies the derivation of a number of important basic relationships, which we shall shortly need to establish. The function $f(x)$ is called the frequency function (f.f.) of the variate X.

Discrete variates

Variates X of the kind just described are such that only certain distinct values have non-zero values of $f(x)$. Variates of this type are said to be discrete variates.

Histograms

We have already pointed out that two kinds of variate arise in practice. With regard to the definition of the f.f. of the second variate type, it is helpful to consider a particular diagrammatic representation of the f.f. of a discrete variate. Take that of (2.10), we

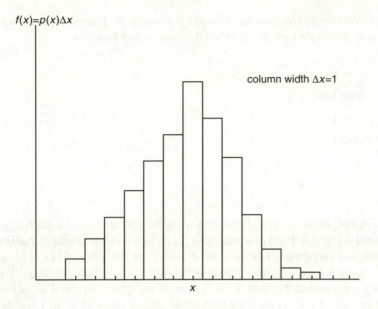

$f(x)=p(x)\Delta x$

column width $\Delta x=1$

x

Fig. 2.3   Histogram.

can illustrate its f.f. by constructing a diagram with columns whose areas are equal to $p(x)$. If these, centred on the values $1, 2, \ldots, n$ are of unit width, the resulting diagram is illustrated by Fig. 2.3. Not all discrete variates take only integer values, the proportion of defective items in an n-fold sample has values $0, 1/n, 2/n, \ldots, 1$. We can represent the probability that X takes one of these values by drawing a histogram with column widths Δx, and height $f(x)$, with $f(x)\Delta x = p(x)$ where $\Delta x = 1/n$. The area of the column of a histogram is the probability that a variate takes a particular value.

To formulate decision rules we need the probabilities that variate X is greater than a specified value x_1. For discrete variates these total probabilities are obtained by summing individual probabilities $p(x)$ for $x > x_1$. We shall see that for continuous variates the probability that X takes a particular value is 0 and accordingly the concept of the f.f. of a continuous variate needs further consideration. The formulation of decision rules both for discrete and continuous variates, requires the definition of a new function which is related to $f(x)$. Consider a discrete variate which only takes a succession of integer values, so that $p(x) = f(x)$. The probability that $X \leq x_1$ is the sum of all the values of $f(x)$ for all the values of $x \leq x_1$. If we denote this probability by $F(x_1)$ then

$$F(x_1) = \sum_{x=-\infty}^{x=x_1} f(x).$$

$F(x)$ is called the distribution function (d.f.) of variate X. The rectangular distribution defined by the single toss of a dice has the distribution function

$$
\begin{aligned}
&= 0 && x < 1 \\
&= 1/6 && 1 \le x < 2 \\
&= 1/3 && 2 \le x < 3 \\
F(x) &= 1/2 && 3 \le x < 4 \\
&\quad \cdots \\
&= 1 && 6 \le x.
\end{aligned}
$$

An example of the general shape of the distribution function of a discrete variate is illustrated in Fig. 2.4. It is a discontinuous function with several simple, but important mathematical properties. Let us write $F(-\infty)$ for the value $F(x)$ approaches as $x \to -\infty$, and $F(\infty)$ for the value to which it tends as x becomes large. Evidently $F(-\infty)$ is the probability that X takes none of its values, so that $F(-\infty) = 0$. On the other hand, $F(\infty)$ is the total probability that X takes all of its possible values and is therefore equal to 1.

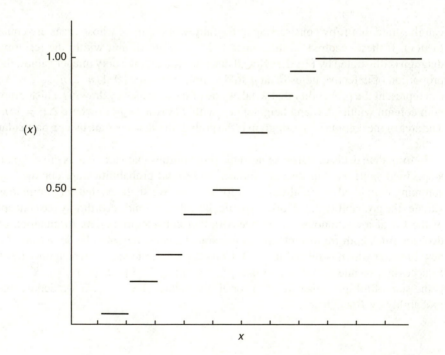

Fig. 2.4   Distribution frequency of a discrete variate.

For particular values x_1 and x_2 of X with $x_2 > x_1$,

$$F(x_2) - F(x_1) = p(x_1 < X \leq x_2),$$

so that $F(x_1) \leq F(x_2)$, and since $0 \leq p(x_1 < X \leq x_2)$, $F(x)$ is a discontinuous function of x which lies between 0 and 1, and which increases in value as x increases. In formal mathematical language it is a right continuous bounded monotonic increasing function of x! Its definition leads to probability statements, which can be used to formulate statistical decision rules for assessing the output of processes where measures of quality are continuous variates.

Continuous variates

Histograms can be used as a convenient visual aid to define the frequency function of a continuous variate. Those of Fig 2.5, illustrate typical changes in the shape of a histogram, when it is used to summarize a large amount of data for a continuous variate and Δx becomes smaller. The data could be measurements of the heights of male adults in a very large population. For the histogram, variate values X are grouped into ranges of values with width Δx, X can take any value between 3 ft and 8 ft (say). To take another example, we could record the lengths L of telephone calls of less than 10 minutes dealt with by an exchange over a long period of time. L obviously takes any value between 0 and 10 minutes. We could count the number of calls falling in 30-second intervals between 0 and 600 seconds. On the other hand, we could group the calls into 10-second intervals or less. In doing so the histograms would resemble those of Fig 2.5. As the interval lengths become smaller each histogram is a progressive approximation to the statistical distribution of X or L. Consider data grouped into intervals of width Δx, the probability that X will fall in the range between $x - \Delta x/2$ and $x + \Delta x/2$ is,

$$p(x - \Delta x/2 < X \leq x + \Delta x/2) \cong g(x)\Delta x,$$

or, equivalently,

$$p(X \leq x + \Delta x/2) - p(X \leq x - \Delta x/2) \cong g(x)\Delta x. \tag{2.11}$$

As in the case of discrete variates, let us write $p(X \leq x + \Delta x/2)$, the total probability that $X \leq x + \Delta x/2$ as $F(x + \Delta x/2)$. In this notation (2.11) then becomes

$$[F(x + \Delta x/2) - F(x - \Delta x/2)]/\Delta x \cong g(x).$$

Let us now write the limit of $g(x)$ as Δx approaches 0 as $f(x)$, so that

$$\lim_{\Delta x \to 0} g(x) = f(x)$$

and

$$\lim_{\Delta x \to 0} \{[F(x + \Delta x/2) - F(x - \Delta x/2)]/\Delta x\} = f(x),$$

giving,

$$d\{F(x)\} = f(x)\,dx. \tag{2.12}$$

$F(X)$ is the distribution function of X, so that $F(x) = p(X \leq x)$, and $f(x)$ is its frequency function, also $F(-\infty) = 0$ and $F(\infty) = 1$ and

$$F(x_1) = \int_{-\infty}^{x_1} f(x)\,dx$$

and

$$\int_{-\infty}^{+\infty} f(x)\,dx = 1.$$

It is clear from the derivation of $f(x)$, that we can only speak of the probability that a continuous variate X takes a particular value x in a mathematical sense; namely, that it is $f(x)\,dx$. For a particular histogram, the probability that X lies in the interval $x - (\Delta x/2)$ to $x + (\Delta x/2)$ is $g(x)\Delta x$. This product obviously approaches 0 as Δx becomes smaller. Thus the only non-zero probabilities we can evaluate for continuous variates are total probabilities. Examples are,

$$p(X \leq x_1) = \int_{-\infty}^{x_1} f(x)\,dx = F(x_1)$$

$$p(X \geq x_1) = \int_{x_1}^{\infty} f(x)\,dx = 1 - F(x_1)$$

$$p(x_1 < X \leq x_2) = \int_{x_1}^{x_2} f(x)\,dx = F(x_2) - F(x_1).$$

Evidently any function $f(x)$ is a f.f. if

- $f(x) \geq 0$ for all x in the range $-\infty \leq x \leq \infty$;
- either $\sum_{x=-\infty}^{x=\infty} f(x) = 1$ or $\int_{x=-\infty}^{x=\infty} f(x)\,dx = 1$.

The simplest example of a continuous variate is that of the rectangular variate illustrated in Fig. 2.2, namely,

$$f(x) = 1/2a \quad -a < x \leq a$$
$$= 0 \quad \text{otherwise.}$$

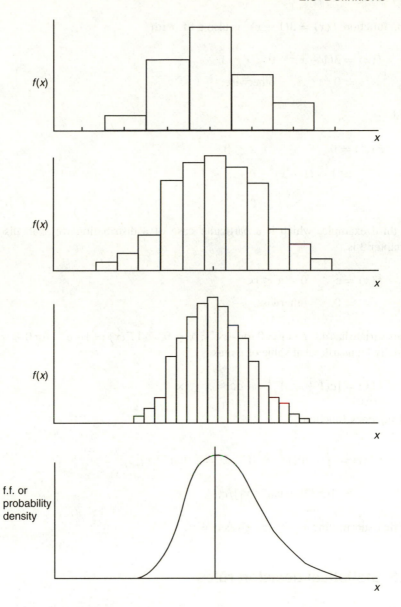

Fig. 2.5 Progressive approach of a histogram to a continuous curve.

Since $f(x) = 0$ for $X \leq -a$ and for $X > a$ we have

$$F(x) = 0 \qquad\qquad x \leq -a$$
$$= (x + a)/2a \quad -a < x \leq a$$
$$= 1 \qquad\qquad x > a.$$

The function $f(x) = 3(1 - x)^2$ is also a f.f. with

$$f(x) = 3(1 - x)^2 \quad 0 < x \leq 1$$
$$= 0 \qquad\qquad \text{otherwise,}$$

and

$$F(X) = 0 \qquad\qquad\qquad x \leq 0$$
$$= 1 - (1 - x)^3 \quad 0 < x \leq 1$$
$$= 1 \qquad\qquad\qquad x > 1.$$

A third example, which is a particular case of a distribution we shall discuss in Chapter 3 is

$$f(x) = e^{-x} \quad 0 < x \leq \infty$$
$$= 0 \qquad \text{otherwise.}$$

This variate has d.f. $F(x) = 0$ for $-\infty < x \leq 0$, and $F(x) = 1 - e^{-x}$ for $0 < x \leq \infty$. Finally let us ask what value of c makes

$$f(x) = [c(1 + x^2)]^{-1} \quad -\infty < x \leq \infty$$

a frequency function?

$$F(x_1) = \int_{-\infty}^{x_1} [c(1 + x^2)]^{-1} \, dx = [c \tan^{-1} x]_{-\infty}^{x_1}$$
$$= [(\pi/2) + \tan^{-1} x_1]/c,$$

so that substituting $x_1 = -\infty$ gives $c = \pi$.

2.6 Statistical decision rules

Using the frequency function derived for the occurrence of x defectives in a random sample of n articles from a population with proportion p defectives, we can return to the example discussed in Chapter 1. We have indicated that use of a rule based on whether or not x exceeds k can result in two kinds of wrong decision. We can decide that a batch of material is rejectable when it is not, and we can accept a batch of material when it should be rejected. We cannot avoid reaching incorrect decisions; however, we can control the frequency with which they occur. To illustrate this is so, let us use the following notation and write $f(x; n, p)$ for the f.f. of a binomial variate X with parameters n and p, so that

$$f(x; n, p) = {}^nC_r p^x q^{n-x} \qquad x = 1, 2, 3, \ldots, n$$

$$= 0 \qquad\qquad \text{otherwise.}$$

From eqn (2.10) it therefore follows that the probability of k or more defectives in a sample of n is

$$Pr(k; n, p) = \sum_{x=k}^{x=n} f(x; n, p) = 1 - F(k - 1; n, p).$$

$Pr(k; n, p_a)$ is then the probability of k or more defectives when a process is running at AQL, and $F(k - 1; n, p_r)$ is the corresponding probability at RQL.

Table 2.1 gives values of $f(x; n, p)$ and $F(x; n, p)$ for $p_a = 0.05$, $p_r = 0.15$ and $n = 50$. From it we see that if $k = 5$ the probability of wrongly rejecting batches of material is approximately 10 per cent, since

$$Pr(5; 50, 0.05) = 0.1036$$

whilst the probability of accepting rejectable batches is

$$F(4; 50, 0.15) = 0.1121.$$

Let us examine what happens to $F(k; n, p_a)$ and $F(k; n, p_r)$ if we use different values of n and k. Suppose we want the probability of rejecting acceptable material

Table 2.1 Values of $f(x)$ and $F(x)$ for $n = 50$, $p = 0.05$ and $p = 0.15$

	$p = 0.05$		$p = 0.15$	
x	$f(x)$	$F(x)$	$f(x)$	$F(x)$
0	0.07694	0.07694	0.00030	0.00030
1	0.20249	0.27943	0.00261	0.00291
2	0.26110	0.54053	0.01128	0.01419
3	0.21988	0.76041	0.03186	0.04605
4	0.13597	0.89638	0.06606	0.11211
5	0.06584	0.96222	0.10724	0.21935
6	0.02602	0.98824	0.14195	0.36130
7	0.00857	0.99681	0.15745	0.51875
8	0.00243	0.99924	0.14935	0.66810
9	0.00060	0.99984	0.12299	0.79109
10	0 00013	0.99997	0.09677	0.88008
11	0.00003	1.00000	0.08819	0.93719
12			0.05711	0.96994

Table 2.2   Values of $F(x)$ when $p = 0.05$ and 0.15 with $n = 70$ and 100

	n = 70			n = 100	
	p = 0.05	p = 0.15		p = 0.05	p = 0.15
x	F(x)	F(x)	x	F(x)	F(x)
5	0.86277	0.03840	8	0.93691	0.02748
6	0.93965	0.08380	9	0.97181	0.05509
7	0.97664	0.15704	10	0.98853	0.09945
8	0.99197	0.25883	11	0.99573	0.16349
			12	0.99584	0.24730

to be less than 5 per cent, can we find a rule with this property? If we take $n = 70$ we see from Table 2.2 that when $k = 8$,

$$Pr(8; 70, 0.05) = 0.0234 \quad \text{and} \quad F(7; 70, 0.15) = 0.1570.$$

This new rule reduces the chance of wrong rejection of quality to less than 5 per cent at the expense of increasing the chance of wrongly accepting rejectable material. Do we have to compromise with regard to probabilities of wrong decisions, in the sense of an increase in one being the price for a decrease in the other? Examination of Table 2.2 indicates that we may be able to achieve specific risks of wrong decisions when $p = p_a$ and $p = p_r$. Thus, suppose we want a rule with a chance of 5 per cent or less of rejecting material with $p \leq 0.05$, and a chance of 10 per cent or less of accepting material when $p \geq 0.15$. If we increase n to 70 and take $k = 7$ rather than 8, the risk of wrong rejection is 6.4 per cent whilst that of accepting rejectable material is 8.4 per cent. These values suggest that we should be able to find n and k which meet both risks. Suppose we want a rule with both chances of wrong decisions ≤ 0.05. It is clear that a rule with this feature is obtained when $k = 10$ for then

$$Pr(10; 100, 0.05) = 0.0282 \quad \text{and} \quad F(9; 100, 0.15) = 0.0551.$$

From a practical point of view this latest rule has more realistic properties than its predecessors, since it ensures that the probabilities of either wrong decision is small.

We shall return to the design of sound decision rules when we have developed some basic theory which we need. It is sufficient, for the moment, to emphasise that probabilities of realistic magnitude were achieved by changing,

- the criterion on which the decision was based; that is, the value of k;
- the size of the sample.

The examples examined indicate that there may be a whole variety of schemes with similar probabilities of drawing wrong conclusions at AQL and RQL. If this is so, are some schemes better than others? If so, in what sense are they better? Much of the methods of statistics can be understood by carrying through calculations of the kind illustrated in Tables 2.1 and 2.2. The arithmetical effort required to do so, however, looks somewhat awesome! To control the frequencies of incorrect decisions at values α and β (say) we need to determine values of n and k which satisfy the equations.

$$Pr(k; n, p) = \sum_{x=k}^{x=n} {}^nC_x p^x q^{n-x} = \alpha \quad \text{when } p = p_a \tag{2.13}$$

and

$$1 - F(k; n, p) = \sum_{x=0}^{x=k-1} {}^nC_x p^x q^{n-x} = \beta \quad \text{when } p = p_r. \tag{2.14}$$

Unless we can obtain appropriate formal approximations for the summations in these expressions, the determination of the values of n and k and study of the properties of schemes based on them will be cumbersome, difficult, and time-consuming. Adequate approximations to values of $F(x; n, p)$ are clearly essential, if a wide range of decision rules are to be assessed with relative ease.

2.7 Summarizing parameters

How should we set about the task of formulating algebraic expressions which approximate frequency functions like,

$$f(x; n, p) = {}^nC_x p^x q^{n-x} \quad x = 0, 1, 2, \ldots, n$$
$$= 0 \quad \text{otherwise,} \tag{2.15}$$

which are reasonably straightforward to manipulate mathematically?

Many statistical studies are based on data sets obtained from experimentation or observation which we need to generalize in order to draw conclusions from them. An example is large clinical trials to assess the effectiveness of different drugs in the treatment of a particular disease. Another relates to the quality of testing in a hospital laboratory or the manufacture of medicinal products by a pharmaceutical company. In such circumstances we often need to periodically check the performance of various dispensing and testing equipment to ensure the precision of results obtained with their use. A further clinical example is the determination of the concentration of specific constituents in patient samples. When several methods of analysis are available to do so, we may want to know which one gives the most accurate results. To answer questions of this kind, we need to express basic data in formal terms amenable to relatively straightforward mathematical methods. To begin the task of formally summarizing collected data, or seeking approximations to frequency functions like (2.15) consider the histograms shown in Fig. 2.2. We obviously need to locate the position of a central histogram value on the line representing values of X, and in addition we need a measure of its width.

Expected value

A parameter which is used to locate a distribution of practical data or that of a theoretical f.f., $f(x)$ is the average value of variate X. We call this average the mean or
expected value of X. This is the sum of all of the values that X takes, weighted by
its frequency of occurrence. Thus for a discrete variate X with f.f. $f(x)$, its expected
value $E(X)$ is

$$E(X) = \sum_{-\infty}^{+\infty} xf(x).$$

The corresponding expression when X is a continuous variate is

$$E(X) = \int_{-\infty}^{+\infty} xf(x)\,\mathrm{d}x.$$

The expected value of a function of $\alpha(X)$ of X, denoted by $E[\alpha(X)]$ is

$$E[\alpha(X)] = \sum_{-\infty}^{+\infty} \alpha(x)f(x) \text{ or } = \int_{-\infty}^{+\infty} \alpha(x)f(x)\,\mathrm{d}x,$$

depending on whether X is discrete or continuous. Thus if $\alpha(X) = X^2$ and X is
continuous then

$$E(X^2) = \int_{-\infty}^{+\infty} x^2 f(x)\,\mathrm{d}x.$$

The mean of a variate is often denoted by m, so that for a continuous variate

$$E(X) = m = \int_{-\infty}^{+\infty} xf(x)\,\mathrm{d}x.$$

For distribution width, we need a measure which indicates its spread on either side
of the mean and the extent to which, as measured by $f(x)$, values of X cluster round
this value. To achieve this objective we clearly need sums of deviations of X from its
mean m weighted by $f(x)$. One such measure is obtained if we use $\alpha(X) = (X-m)^2$.
This expected value defines an important distribution parameter called the variance
$V(X)$ of variate X, so that

$$V(X) = \sum_{-\infty}^{+\infty} (x - m)^2 f(x) \quad \text{or} \quad = \int_{-\infty}^{+\infty} (x - m)^2 f(x)\,\mathrm{d}x,$$

depending on whether X is discrete or continuous.

The square root of $V(X)$, usually denoted by $\sigma(X)$, is called the standard deviation of X. To calculate, or obtain a formal expression for the variance of X it is often more convenient to use the equation (2.16) below. Thus if X is discrete we have,

$$V(X) = \sum_{-\infty}^{+\infty}(x^2 - 2mx + m^2)f(x) = \sum_{-\infty}^{+\infty}x^2 f(x) - 2m\sum_{-\infty}^{+\infty}xf(x) + m^2$$

$$= E(X^2) - m^2. \tag{2.16}$$

The same result obviously holds when X is a continuous variate.

Let us consider some examples which illustrate the definitions we have formulated in this chapter and which relate to matters we shall discuss in Chapter 3.

Example 2.1

The continuous variate X has f.f. defined by

$$f(x) = ce^{-2x} \quad 0 < x \le \infty$$
$$= 0 \quad \text{otherwise,}$$

what is the value of c? Obtain the probability that $X \le 1$ and the probability that X lies in the range defined by $0.25 < X \le 0.75$. What is the value of x for which $p(X \ge x)$ is 0.05? Obtain the mean and standard deviation of X.

Solution

$$F(x_1) = c\int_0^{x_1} e^{-2x}\, dx \quad x_1 > 0$$

$$= c(1 - e^{-2x_1})/2$$

$$F(x_1) = 0, \quad x_1 \le 0$$

since $F(\infty) = 1$, $c = 2$ and $p(X \le 1) = (1 - e^{-2}) = 0.8647$, whilst

$$p(0.25 < X \le 0.75) = F(0.75) - F(0.25) = 0.3834.$$

To obtain the value of x_1 for which $p(X \ge x_1) = 0.05$, we have

$$p(X \ge x_1) = 1 - F(x_1) = e^{-2x_1}$$

so that $2x_1 = -\ln(0.05)$ giving $x_1 = 1.4979$. Finally we have,

$$E(X) = \int_0^\infty 2xe^{-2x}\, dx = \int_0^\infty e^{-2x}\, dx = 1/2$$

and

$$E(X^2) = 2 \int_0^\infty x^2 e^{-2x} \, dx = 2 \int_0^\infty x e^{-2x} \, dx = 1/2,$$

which gives $V(X) = 1/4$ and $\sigma(X) = 1/2$. Notice that for this distribution m and $\sigma(X)$ are equal to each other.

Example 2.2

A variate X has f.f.

$$f(x) = 2/(1+x)^3 \quad 0 < x \le \infty$$
$$= 0 \qquad\qquad \text{otherwise.}$$

If a sample of 10 independent values of X is taken from this population, what is the probability that two of them will be ≥ 1?

Solution

Since $F(x) = 0$ for $x < 0$,

$$p(X \ge 1) = 2 \int_1^\infty (1+x)^{-3} \, dx = \left[-(1+x)^{-2} \right]_1^\infty = 0.25 = p_1.$$

From eqn (2. 10), and using the notation of Section 2.3, the probability of two samples with $X \ge 1$ is therefore,

$$f(2;\ 10,\ p_1) = 45.(0.25)^2 (0.75)^8 = 0.2816$$

Example 2.3

From Table 2. 1 calculate the mean number of defective items in a sample of 50 items taken from a process, when the proportion of such items is $p_r = 0.15$. Also obtain their standard deviation and confirm that the probability of obtaining a number of defective items outside the range $m \pm 2\sigma(X)$, is about 1 in 23.

Solution

We find $m = 7.5$, $\sigma(X) = 2.525$ so that $m + 2\sigma(X) = 12.55$, $m - 2\sigma(X) = 2.45$ whilst from the table we see that $p(X \ge 13) = 0.0301$, $p(X \le 2) = 0.0142$ giving $p(X \le 2,\ \text{or}\ X \ge 13) = 0.0443$, that is, odds of 1 in 22.6.

Let us speculate about $\sigma(X)$ since it is a measure of the width of the distribution of X, a decrease in $\sigma(X)$ suggests that the distribution of X about its mean m becomes narrower, so that the probability that it lies closer to m increases. A question now arises, namely, is there a relationship between the mean and standard deviation of a variate and probability statements which can be made about it? A result which indicates this is a strong possibility is easily established. Suppose X is a continuous variate with f.f. $f(x)$, mean m and finite standard deviation σ, so that

$$\sigma^2 = \int_{-\infty}^{\infty} (x - m)^2 f(x)\,dx.$$

For any value of $k > 0$ we can write

$$\sigma^2 = \int_{-\infty}^{m-k\sigma} (x-m)^2 f(x)\,dx + \int_{m-k\sigma}^{m+k\sigma} (x-m)^2 f(x)\,dx + \int_{m+k\sigma}^{\infty} (x-m)^2 f(x)\,dx$$

and

$$\sigma^2 \geq \int_{-\infty}^{m-k\sigma} (x - m)^2 f(x)\,dx + \int_{m+k\sigma}^{\infty} (x - m)^2 f(x)\,dx.$$

For the first integral in this equation $x \leq m - k\sigma$ which gives $(x - m) \leq -k\sigma$ and $(x - m)^2 \geq k^2\sigma^2$ therefore,

$$\int_{-\infty}^{m-k\sigma} (x - m)^2 f(x)\,dx \geq k^2\sigma^2 \int_{-\infty}^{m-k\sigma} f(x)\,dx;$$

similarly,

$$\int_{m+k\sigma}^{\infty} (x - m)^2 f(x)\,dx \geq k^2\sigma^2 \int_{m+k\sigma}^{\infty} f(x)\,dx$$

it follows that,

$$\int_{-\infty}^{m-k\sigma} f(x)\,dx + \int_{m+k\sigma}^{\infty} f(x)\,dx = p(|X - m| \geq k\sigma) \leq 1/k^2 \tag{2.17}$$

We therefore obtain the result that the probability that a randomly selected variate differs from its mean m by more than k times its standard deviation is less than or equal $1/k^2$. Thus the probability that X will differ from its mean by more than 3 is $\leq 1/9$. This theorem, due to Bienayme in 1853 and more generally by Tchebycheff, is of no great practical use, since the actual probabilities are usually much smaller than the value given by the theorem. It does, however, encourage us to look seriously for an approximating function, based on m and σ, to replace mathematically awkward summations such as those of the f.f. of the kind given by (2.16). We shall return to this objective in Chapter 3.

3
Distributions relevant to process and test control

The distributions which play an important role in the statistical theory of quality control are the binomial, Poisson, normal, and gamma distributions. We begin with their definitions and properties which are relevant to it.

3.1 Binomial distribution

We have already seen that this is a distribution which is associated with situation where there are just two outcomes when a single sample is drawn from a population. The variate X then takes the values 0 or 1 with probabilities q and p. If the two outcomes, or events, are A and not A, give the score 1 to the sample result A and 0 to not A, so that $E(X) = p$, $E(X^2) = p$ and from eqn (2.16) $V(X) = pq$. For a sample of n from this population the f.f. of X in the notation of Chapter 2 is $f(x; n, p)$, and

$$E(X) = \sum_{x=0}^{x=n} x f(x; n, p) = \sum_{x=1}^{x=n} {}^n C_{x-1} p^x q^{n-x}$$

$$= np \sum_{x=1}^{x=n} {}^{n-1} C_{x-1} p^{x-1} q^{n-x}$$

writing $y = x - 1$ gives

$$E(X) = np \sum_{y=0}^{y=n-1} {}^{n-1} C_y p^y q^{n-1-y} = np(p+q)^{n-1} = np. \tag{3.1}$$

Using the same procedure, and the relationship $x^{(2)} = x(x-1)$ so that

$$E(X^2) = \sum_{x=0}^{x=n} x^{(2)} f(x; n, p) + \sum_{x=0}^{x=n} x f(x; n, p)$$

we find

$$E(X^2) = n^{(2)} p^2 + np.$$

From (2.16) it follows that

$$V(X) = npq \quad \text{and} \quad \sigma(X) = \sqrt{(npq)}. \tag{3.2}$$

Let us consider the f.f. of X in a little more detail, we have

$$f(x; n, p)/f(x - 1; n, p) = (n - x + 1)p/xq = 1 + [(n + 1)p - x]/xq.$$

$f(x; n, p)$ is therefore a steadily increasing function of x as long as $(n + 1)p \geq x$ and steadily decreases thereafter. If x_c is an integer such that

$$(n + 1)p - 1 < x_c \leq (n + 1)p,$$

then x_c is the most probable number of successes in a sample of n, $f(x_c, n, p)$ is called the central term of the distribution. If $(n+1)p$ is an integer, then $f(x_c; n, p) = f(x_c - 1; n, p)$. It follows from the above ratio that for $x \geq (r + 1)$ and $r \geq x_c$

$$f(x; n, p)/f(x - 1; n, p) \leq (n - r)p/(r + 1)q$$

so that

$$f(r + \omega; n, p)/f(r; n, p) \leq [(n - r)p/(r + 1)q]^\omega$$

and so for $r > np$

$$\sum_{\omega=0}^{\omega=n-r} f(r + \omega; n, p) \leq f(r; n, p) \sum_{\omega=0}^{\omega=n-r} [(n - r)p/(r + 1)q]^\omega$$

$$\leq f(r; n, p) \sum_{\omega=0}^{\omega=\infty} [(n - r)p/(r + 1)q]^\omega$$

$$= f(r; n, p)(r + 1)q/[(r + 1) - (n + 1)p]$$

$$< f(r; n, p)rq/(r - np).$$

There are at least $(r - np)$ values of $f(x; n, p)$ greater than $f(r; n, p)$; these sum to less than 1, so that $f(r; n, p)$ is less than $1/(r - np)$. We therefore conclude that for the probability $p(X \geq r)$ that $X \geq r$,

$$p(X \geq r) \leq rq/(r - np)^2. \tag{3.3}$$

For $r < np$ similar reasoning leads to,

$$p(X < r) \leq (n - r)p/(np - r)^2. \tag{3.4}$$

The inequalities (3.3) and (3.4), although crude are important, since they permit a simple formalization of what has so far been an entirely intuitive assumption of the notion of probability. Suppose in n identical and independent trials, the event E occurs r times. If the probability of E is p then, when n is large we intuitively expect r/n to be close to p. This assumption is, of course, the basis of sampling inspection schemes for the control of the proportion of defective items. If X is the number of successes in n binomial trials with probability of success p, consider the probability that $(X/n) > p + \xi$ where $\xi > 0$ is fixed and small. Clearly, $p[(X/n) > p + \xi]$ is equal to $p[(X > n(p + \xi)]$ and so from (3.3),

$$p[X > n(p + \xi)] \leq 1/n\xi^2.$$

Thus, as n increases $p[X > n(p + \xi)] \to 0$, and use of (3.4) likewise gives the result $p[X < n(p - \xi)] \to 0$. We therefore have that, as n increases so

$$p[|(X/n) - p| < \xi] \to 1,$$

and we have the following theorem

Theorem

In a binomial situation with probability of success p, the probability that the average number of successes obtained in an n-fold sample deviates from p by more than a specified but small quantity ξ tends to zero as n increases.

This assertion is an example of what is known as the weak law of large numbers. It is of considerable theoretical importance, notwithstanding its theoretical significance, it is far too weak to be of practical interest. In formal terms we summarize the above theorem by saying that (X/n) converges in probability to p. Evidently for any constant C, $E[(CX)^r] = C^r E(X^r)$ so that from (3.1) and (3.2),

$$E(X/n) = p \quad \text{and} \quad V(X/n) = pq/n.$$

We can use these two expectations and (2.17) to establish the same result, for we have

$$p[|(X/n) - p| \geq k\sqrt{(pq/n)}] \leq 1/k^2.$$

If in this expression we write ξ for $k\sqrt{(pq/n)}$, then we can make the probability that X/n deviates from p as small as we like by making n sufficiently large.

3.2 The moments of a distribution

For a variate X with f.f. $f(x)$ and mean m we define its rth zero moment as

$$\mu_r(X) = \sum_{-\infty}^{\infty} x^r f(x) \quad \text{or} \quad \int_{-\infty}^{\infty} x^r f(x)\, \mathrm{d}x,$$

depending on whether X is discrete or continuous. Its rth mean moment is,

$$m_r(X) = \sum_{-\infty}^{\infty} (x - m)^r f(x) \quad \text{or} \quad \int_{-\infty}^{\infty} (x - m)^r f(x)\, dx$$

The moments of a distribution give us information about its shape and the extent to which one distribution differs from another. For the kinds of distribution which arise in practice, it is generally true, that the more equivalent moments two frequency functions $f_1(x)$ and $f_2(x)$ have, the closer they are in shape. Thus, if two distributions have a number of moments in common we can envisage using the f.f. of one to replace the other in the calculation of total probabilities. We can indicate that this is so, if we consider approximating a frequency function with a polynomial $g(x)$. Suppose $f_1(x)$ and $f_2(x)$ are continuous frequency functions both with range $x = a$ to $x = b$ and with equivalent moments to order k. Take the f.f. $f_1(x)$ and fit a polynomial $g(x)$ of degree k to it by the method of least squares so that

$$g(x) = \sum_{r=0}^{r=k} \alpha_r x^r$$

and we determine α_r such that

$$\int_a^b [f_1(x) - g(x)]^2\, dx \tag{3.5}$$

takes its minimum value. Differentiating (3.5) with respect to α_i $(i = 1 \text{ to } k)$ gives,

$$\int_a^b x^i f_1(x)\, dx = \mu_i(x) = \int_a^b \sum_{r=0}^{r=k} \alpha_r x^{r+i}\, dx. \tag{3.6}$$

The set of equations generated by (3.6) clearly yield the same polynomial for both $f_1(x)$ and $f_2(x)$, so that if

$$|f_1(x) - g(x)| = \xi_1(x) \quad \text{and} \quad |f_2(x) - g(x)| = \xi_2(x)$$

then

$$|f_1(x) - f_2(x)| \le \xi_1(x) + \xi_2(x),$$

where $\xi_1(x)$ and $\xi_2(x)$ are small. Generally speaking the more moments $f_1(x)$ and $f_2(x)$ have in common (that is, the higher the value of k) the closer $g(x)$ will approximate to $f_1(x)$ and $f_2(x)$.

Let us consider one or two examples to illustrate aspects of binomial variates which we have developed.

Example 3.1

A machine produces a large number of items automatically. When it is in control it is known that 5 per cent of its output is defective. Samples of 50 components are taken at a time. What is the probability that

- such a sample contains at least 5 defectives;
- 3 successive samples contain at least 5 defectives;
- each of 4 samples contain just 2 defectives;
- the total number of defective items in two successive samples is 5?

Solution

The probability of x defectives in a sample of n is

$$f(x; n, p) = {}^nC_x p^x q^{n-x}$$

so that the probability of at least 5 defective items is

$$1 - \sum_{x=0}^{x=4} {}^{50}C_x (0.05)^x (0.95)^{50-x}.$$

Now

$$f(0; 50, 0.05) = (0.95)^{50} = 0.0769$$

and so, using the generating equation of Section 3.1, namely,

$$f(x; n, p) = f(x - 1; n, p)(n - x + 1)p/xq, \tag{3.7}$$

we find

$$f(1; 50, 0.05) = 0.2025.$$

Repeated use of (3.7) then gives $p(X \geq 5) = 0.1036$. The probability of 3 successive samples each having 5 or more defectives is $p(X \geq 5)^3 = 0.0011$.

Two defective items in a sample of 50 occur with frequency $f(2; 52, 0.05)$ which is 0.2611 so that 4 successive samples with 2 defectives will have probability $[f(2; 50, 0.05)]^4 = 0.0046$.

The probability that the total number of defectives in 2 successive samples 5 is

$$\sum_{x=0}^{x=5} f(x; 50; 50, 0.05) \cdot f(5 - x; 50, 0.05) = 0.1800.$$

Example 3.2

Metal bars produced using a particular process were tested to examine the distribution of weak spots along their lengths. Each of 300 bars was put under stress in 6 places, and the number of places where each failed was recorded. The data obtained is given in the table below. In theory the failures should be evenly distributed along the bar. Does the data indicate that this is so?

Number of failures x	Number of bars with x failures
0	132
1	119
2	39
3	6
4	3
5	1

Solution

From the data we see that the total number of failures observed was 232, so that the average number for a single bar was 0.7733. If the failures are evenly distributed along a bar, then X is a binomial variate. Thus, from (3.1) if p is the probability of a failure then the average number of failures for n tests on a bar is np. Therefore, in this example if we write p_e for an estimate of p, we can write $6p_e = 0.7733$ and $p_e = 0.1289$. Using this value for p the expected number of bars with x failures is then $300 f(x; 6, 0.1289)$. The values of this expression for $x = 0$ to 6 are shown in the table below. From them it looks as though failures are indeed evenly distributed along the bar lengths.

Number of failures x	Predicted number of failures
0	131
1	116
2	43
3	9
4	1
5	0

Example 3.3

If X is a binomial variate with parameters n and p use the Tchebycheff inequality to establish that

$$p[|(X/n) - p| \geq \xi] \leq 1/4n\xi^2.$$

Compare the values given by this inequality with those of $p[0.30 < (X/n) < 0.70]$ when $p = 0.5$, $n = 10$, and $n = 50$.

Solution

For $n = 10$

$$p[0.30 < (X/n) < 0.70] = 1 - 2p(X \geq 7)$$

$$p(X \geq 7) = \sum_{x=7}^{x=10} f(x; 10, 0.50) = 0.1719$$

$$\therefore p[0.30 < (X/n) < 0.70] = 0.6562.$$

When $n = 50$

$$p[0.30 < (X/n) < 0.70] = 1 - 2p(X \geq 31).$$

By using the recurrence relationship (3.7) or tables of the binomial d.f., we find that

$$p(X \geq 31) = 0.0595$$

Therefore, $p[0.30 < (X/n) < 0.70] = 08810$.

In the inequality at the end of Section 3.1, put $k = \xi \sqrt{(n/pq)}$; it then follows that

$$p[|(X/n) - p| \geq \xi] \leq pq/n\xi^2 \leq 1/4n\xi^2,$$

and so for $n = 10$ and $\xi = 0.20$,

$$p[|(X/n) - 0.50| \geq 0.20] \leq 0.625.$$

For $n = 50$ $p[|(X/n) - 0.50| \geq 0.20] \leq 0.125$.

As we have already indicated, this is an example of the weak law of large numbers. Is there a strong law of large numbers; that is, a much closer approximating expression to the above and similar probabilities? We can illustrate this possibility by considering an example which is an extension of Example 2.3.

Example 3.4

For binomial variates with $p = 0.50$ and $p = 0.20$, $n = 20$, 50, and 100, obtain the value of x_1, where x_1 is the integer nearest to, but less than $m - 2\sigma(X)$. Also, obtain the value of x_2 which is nearest but greater than $m + 2\sigma(X)$, where m is the mean of X and $\sigma(X)$ is its standard deviation. Either using (3.7) or tables of the binomial d.f., obtain the odds that X lies outside the range $x_1 < X$ and $X < x_2$.

Solution

- $n = 20$ and $p = 0.50$, $m = 10$ and $\sqrt{(npq)} = 2.236$, so that $x_1 = 5$ and $x_2 = 15$, and since $p = 0.50$ $p(X \geq 15) = p(X \leq 5) = 0.0207$, the odds that X does not lie in the range $x_1 < X < x_2$ is 1 in 24;
- $n = 50$, $m = 25$ and $\sqrt{(npq)} = 7.06$ $x_1 = 17$ and $x_2 = 33$, we find $p(X \geq 33)$ is 0.0164 giving odds of 1 in 30 for X outside $x_1 < X < x_2$;

- $n = 100, m = 50$ and $\sqrt{(npq)} = 5, x_1 = 39$ and $x_2 = 61$, which gives $p(X \geq 61)$ equal to 0.01760 and odds that X is in the range specified is 1 in 28;
- for $p = 0.20$ similar calculations give $p(x_1 < X < x_2) = 0.9563$, when n is 20, $p(x_1 < X < x_2) = 0.9507$ with $n = 50$ and $p(x_1 < X < x_2) = 0.9635$ for $n = 100$, the respective odds for X outside $x_1 < X < x_2$ are then 1 in 23, 1 in 20, and 1 in 30.

From this example we see that the odds for all of the above values of X outside the range $m \pm 2\sigma(X)$ are broadly speaking very similar, only varying between 1 in 20 and 1 in 30. These calculations could be indicating the existence of a common approximating function for a wide range of values of n and p involving just two parameters m and σ.

Example 3.5

A sample of n items is taken at regular intervals to check the quality of a manufacturing process. When the number of defectives in the sample exceeds k this could be due to a deterioration in the quality of the product. It is decided that the process will be investigated to examine its quality on the rth occasion that this occurs. If variate Y is the number of samples taken for this to happen obtain its frequency function.

Solution

Let p be the probability that in an n-fold sample the proportion of defectives is greater than k. The first $Y - 1$ samples will have $r - 1$ values $> k$. The probability of this event is

$$f(y - 1; r - 1, p) = {}^{(y-1)}C_{(r-1)}p^{r-1}q^{y-r}$$

so that the f.f. of Y is $pf(y - 1; r - 1, p)$; that is,

$$= {}^{(y-1)}C_{(r-1)}p^{r}q^{y-r} \quad y = r, r + 1, r + 2, \ldots$$

$$= 0 \qquad\qquad\qquad \text{otherwise} \tag{3.8}$$

A distribution with this f.f. is called the negative binomial distribution; this is because it is the expansion of $p^{r}(1 - q)^{-r}$.

The mean of a variate locates its distribution, and its variance is a measure of its width, there are two other useful measures of shape namely skewness and kurtosis.

3.3 Skewness and kurtosis

A variate X with f.f. $f(x)$ which is symmetric about its mean m, has $f(y)$ and $f(-y)$ equal to one another when $Y = X - m$, so that $m_3(X) = 0$. If $f(-y)$ is generally less than $f(y)$, $m_3(X)$ will be positive, whilst if it is greater than $f(y)$ the third mean moment of X is negative. Evidently $m_3(X)$ has properties relevant to measuring the lack of symmetry of a distribution about its mean. If this moment is to be used to compare the skewness of different distributions it needs to be standardized to take account of differences in scale which may exist between them. Since $\sigma(X)$ is a measure of the width of a distribution, a suitable index with which to measure skewness is accordingly

$$\beta_1(X) = m_3(X)/\sigma^3(X). \tag{3.9}$$

The variate $(X - m)/\sigma$ is called a standardized variate. Since all such variates have unit standard deviation we need to formulate an additional measure of width to compare their distributions. An index which springs to mind is the fourth moment of such variates which we shall denote by $\beta_2(X)$, so that

$$\beta_2(X) = m_4(X)/\sigma^4(X). \tag{3.10}$$

In order to formulate expressions for $\beta_1(X)$ and $\beta_2(X)$ it is sometimes useful to use the relationships between the mean and zero moments of variates, thus

$$m_3(X) = E(X - m)^3 = E(X^3) - 3mE(X^2) + 2m^3,$$
$$= \mu_3(X) - 3m\mu_2(X) + 2m^3. \tag{3.11}$$

Similarly, we have

$$m_4(X) = \mu_4(X) - 4m\mu_3(X) + 6m^2\mu_2(X) - 3m^3. \tag{3.12}$$

Using its zero moments we find for a binomial variate X with parameters n and p that

$$\beta_1(X) = (q - p)/\sqrt{(npq)} \tag{3.13}$$

and

$$\beta_2(X) = 3(1 - 1/n) + [(1 - 3pq)/npq]. \tag{3.14}$$

Thus $\beta_1(X) \to 0$ as n increases, which suggests that the distribution becomes more symmetric about its mean np as n gets larger. From (3.14), $\beta_2(X) \to 3$ as n increases, a fact we shall return to when we have established some properties of the normal distribution.

3.4 The Poisson distribution

Situations arise in quality control where the proportion p of defective items is very small. When this is the case the sample size required to detect changes in p increases.

If we have a situation with p small, n is large and $m = np$ of moderate size a convenient approximation to the f.f. of X is the Poisson distribution which we can derive as follows; the probability of no defectives in a sample of n is

$$f(0; n, p) = (1 - p)^n = (1 - m/n)^n$$

if we use the limit

$$\lim_{n \to \infty} (1 - z/n)^n = e^{-z} \quad \text{for } z < n,$$

then for large n

$$f(0; n, p) \cong e^{-m}. \tag{3.15}$$

It therefore follows from (3.7) that

$$f(x; n, p) = \{[np - (x - 1)p]/xq\} f(x - 1; n, p)$$

so that for small p and x

$$f(x; n, p) \cong (m/x) f(x - 1; n, p). \tag{3.16}$$

Combining eqns (3.15) and (3.16) gives

$$f(x; n, x) = f(x; m) = m^x e^{-m}/x!$$

To obtain the moments of a variate which has a Poisson distribution with mean m we have

$$E[x^{(2)}] = m^2 \sum_{x=0}^{x=\infty} m^{x-2}/(x - 2)! = m^2.$$

so that $V(X) = m$ and we see that this distribution has both mean and variance equal to m. Using eqns (3.11) and (3.12) we also obtain

$$m_3(X) = m \quad \text{and} \quad m_4(x) = m(1 + 3m),$$

giving

$$\beta_1(X) = 1\sqrt{m} \quad \text{and} \quad \beta_2(X) = 3 + 1/m.$$

Again, we see the skewness of the distribution approaches 0 as m increases in value whilst $\beta_2(X)$ tends to 3.

As well as being the limiting distribution of a binomial variate when p becomes small, the Poisson distribution gives the frequencies with which random events occur in time or space. Thus if X is the number of events which occur in time 0 to T then under certain assumptions X is a Poisson variate. It is accordingly a distribution which is used in the design of schemes to assess the life expectancy of items such as light bulbs or electronic components. Let us assume that the probability of an event E occurring in a small time interval δt, is independent of the number of events which occur in the time 0 to t and that E is not time-dependent. Assume also that the probability of two events or more arising in time δt is so small that it can be ignored. On these assumptions, if E occurs in unit time with probability p, and $p(x; t)$ is the probability of x events in time 0 to t, then $p(x; t + \delta t)$ is given by

$$p(x; t + \delta t) = p(x; t)[1 - p\delta t] + p\delta t p(x - 1; t)$$

and allowing $\delta t \to 0$ yields

$$\frac{dp}{dt}(x; t) = p[p(x - 1; t) - p(x; t)]. \tag{3.17}$$

If we adopt the convention that $p(-1; t) = 0$, since the probability of this event is clearly 0 we have

$$\frac{dp}{dt}(0; t) = -p(0; t)p$$

and so

$$p(0; t) = Ce^{-pt}$$

Substitution of this result in (3.17) gives

$$\frac{dp}{dt}(1; t) + p(1; t)p = Ce^{-pt}$$

so that

$$p(1; t) = C(pt)e^{-pt}.$$

Inductive reasoning easily then leads to the conclusion that

$$p(x; t) = (pt)^x e^{-pt}/x! \tag{3.18}$$

Example 3.6

The probability that a manufactured item will fail in unit time is p. Obtain the probability that its life will exceed t_1 and determine its expected life.

Solution

From (3.18) the probability of it not failing is e^{-pt}. The probability of failure in time $t, t + \delta t$ is $p\delta t$ so that the f.f. of time to failure T is

$$f(t) = pe^{-pt} \, dt$$

and

$$F(t_1) = \int_0^{t_1} pe^{-pt} \, dt = 1 - e^{-pt_1}.$$

The probability that $T \geq t_1$ is

$$1 - F(t_1) = e^{-pt}$$

whilst the expected value of T is

$$E(T) = \int_0^\infty te^{-pt} \, dt = 1/p.$$

Example 3.7

A manufacturer produces integrated circuits used to assemble a module which is used in the construction of television sets. It is a known feature of the production process that 1 in 50 circuits will be faulty. To overcome this problem until the cause of the fault is found, the following agreement is reached between the manufacturer and his customer. He will provide packs of integrated circuits such that 95 per cent of them on average contain no less than 150 perfect ones. How many circuits should be included in a pack to meet this specification?

Solution

If the number of faulty units supplied in a pack is r we need $150 + r$ to allow for them. To formulate a general solution to this problem, let p be the proportion of faulty circuits produced, and n the number of perfect circuits which are to be provided in a pack. When p is small the probability of x unusable circuits is given by (3.18) with $m = (n + r)p$. The undertaking to provide d per cent or less packs with more than r faulty units requires $p(x > r) \leq d/100$. We therefore need the lowest value of r which satisfies the equation

$$e^{-p(n+r)} \sum_{x=0}^{x=r} [(n + r)p]^x / x! \geq 1 - d/100$$

For our particular example $p = 0.02$, $n = 150$, and $d = 5$ so that r has to satisfy

$$e^{-(3+0.02r)} \sum_{x=0}^{x=r} [(3 + 0.02r)p]^x / x! \geq 0.95.$$

The lowest value of r to do so is 6, for then the left-hand side of the above equation is 0.9601. The packs of circuits supplied would therefore contain 156 units.

Example 3.8

It is suggested that the occurrence of a particular fault found in a machine knitted garment may follow a Poisson distribution. The number of faults found in 500 garments was as follows

Number of faults	Number observed
0	70
1	141
2	126
3	84
4	48
5	23
6	6
7	2

By calculating the mean and variance of the number of faults per garment examine whether this assumption is likely to be valid. What is the expected distribution of faults in 500 garments on this assumption?

Solution

From the observed table of faults we find their mean is 2.00 and their variance is 2.42. Since these two values are close to one another the Poisson assumption may well be valid. If we take the mean number of faults to be 2 then the assumption that they follow a Poisson distribution predicts that 67 garments would have no faults, 135 would have one, 135 two, 90 three, 45 four, 23 five, and those with six or more would be 8.

3.5 Useful relationships

Before considering other distributions which we use in the statistics of quality control, it is necessary to establish some relationships which we shall need.

1. Consider a function $f(y)$ which is continuous, and not constant, between two values of y, namely a and b ($b > a$). Let $f_m(y)$ be the minimum value of $f(y)$ in

the range $a \leq y \leq b$ and $f_M(y)$ its maximum value. If I is the integral

$$I = \int_a^b f(y) \, dy,$$

then

$$I > (b-a) f_m(y) \quad \text{and} \quad I < (b-a) f_M(y).$$

It therefore follows that

$$I = (b-a) f(a + \theta b) \quad \text{where } 0 < \theta < 1. \tag{3.19}$$

For two continuous functions $g(y)$ and $f(y)$ between $y = a$ and $y = b$

$$\int_a^b f(y) g(y) \, dy > f_m(y) \int_a^b g(y) \, dy$$

and

$$\int_a^b f(y) g(y) \, dy < f_M(y) \int_a^b g(y) \, dy,$$

so that

$$\int_a^b f(y) g(y) \, dy = f(a + \theta b) \int_a^b g(y) \, dy \quad (0 < \theta < 1). \tag{3.20}$$

2. Suppose the first three derivatives $f^{(1)}(y)$, $f^{(2)}(y)$, and $f^{(3)}(y)$ of $f(y)$ exist and are continuous in the range $y = a$ to $y = b$, then for y, and $y + \alpha$ in this range

$$f(y + \alpha) - f(y) = \int_0^y f^{(1)}(y + \alpha - \omega) \, d\omega$$

$$= y f^{(1)}(\alpha) + \int_0^y \omega f^{(2)}(y + \alpha - \omega) \, d\omega$$

$$= y f^{(1)}(\alpha) + y^2 f^{(2)}(\alpha)/2! + \int_0^y [\omega^2 f^{(3)}(y + \alpha - \omega)/2!] \, d\omega.$$

Using eqn (3.20) we therefore have

$$f(y + \alpha) = f(\alpha) + y f^{(1)}(\alpha) + y^2 f^{(2)}(\alpha)/2!$$
$$+ y^3 f^{(3)}(\alpha + \theta y)/3! \quad (0 < \theta < 1). \tag{3.21}$$

3. As n becomes larger an approximation, known as Stirling's approximation for $n!$ is

$$n! = [\sqrt{(2\pi)}] n^{n+1/2} e^{-n}. \tag{3.22}$$

Table 3.1 Stirling's approximation to $n!$

n	$n!$	Approximation	Ratio
10	$3.62880 \cdot 10^6$	$3.59870 \cdot 10^6$	1.0083
20	$2.43290 \cdot 10^{18}$	$2.42279 \cdot 10^{18}$	1.0042
30	$2.64517 \cdot 10^{32}$	$2.65253 \cdot 10^{32}$	1.0028

It is important to ask for what values of n can we justify using the right-hand side of this equation to replace its left-hand side? Some idea of the level of error introduced into expressions by using this approximation can be obtained from Table 3.1.

It is apparent that the approximation works well for relatively small values of n, namely $n \geq 10$.

4. Finally let us obtain the value of the integral

$$I_a = \int_0^a \exp(-x^2/2) \, dx$$

when a becomes infinite so that

$$I_\infty = \int_0^\infty \exp(-x^2/2) \, dx.$$

Evidently,

$$I_a^2 = \int_0^a \int_0^a \exp[-(x^2 + y^2)/2] \, dx \, dy,$$

if we substitute $x = r \cos \theta$ and $y = r \sin \theta$ it is clear from Fig. 3.1 that

$$\int_0^{\pi/2} \int_0^a r \exp(-r^2/2) \, dr \, d\theta < I_a^2 < \int_0^{\pi/2} \int_0^{a\sqrt{2}} r \exp(-r^2/2) \, dr \, d\theta$$

$$\therefore [1 - \exp(-a^2/2)]\pi/2 < I_a^2 < [1 - \exp(-a^2)]\pi/2$$

so that as a becomes large it follows that $I_\infty = \sqrt{(\pi/2)}$.

Example 3.9

For the sum of the binomial frequencies $\sum_{x=0}^{x=r} f(x; n, p)$ obtain the result

$$\sum_{x=0}^{x=r} f(x; n, p) = 1 - \{n!/[r!(n - r - 1)!]\} \int_0^p \omega^r (1 - \omega)^{(n-r-1)} \, d\omega.$$

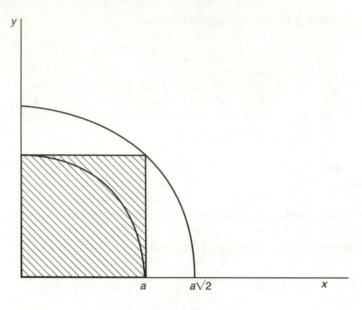

Fig. 3.1

Solution

Assuming that the first $(r+1)$ derivatives of a function $f(y)$ exist and are continuous, repeated application of the procedure which led to (3.21) gives

$$f(y+\alpha) = f(\alpha) + yf^{(1)}(\alpha) + y^2 f^{(2)}(\alpha)/2! + \cdots + y^r f^{(r)}(\alpha)/r!$$
$$+ \{n!/[r!(n-r-1)!]\} \int_0^y \omega^r f^{(r+1)}(y+\alpha-\omega)\, d\omega.$$

Substituting $y = p$, $\alpha = q$ and $f(y+\alpha) = (p+q)^n$ gives

$$(p+q)^n = q^n + npq^{n-1} + n(n-1)p^2 q^{n-2}/2 + \cdots$$
$$+ n(n-1)(n-2)\ldots(n-r-1)p^r q^{n-r} r!$$
$$+ \{n!/[r!(n-r-1)!]\} \int_0^p \omega^r (1-\omega)^{n-r-1}\, d\omega,$$

so that

$$\sum_{x=0}^{x=r} f(x; n, p) = 1 - \{n!/[r!(n-r-1)!]\} \int_0^p \omega^r (1-\omega)^{n-r-1}\, d\omega. \quad (3.23)$$

As a further example, use eqn (3.21) to establish that when $0 < x < 1$,

$$\ln(1+x) = x - x^2/2 + (x^3/3)\delta(x)$$

and,

$$\ln(1 - x) = -x - x^2/2 - (x^3/3)\delta(x),$$

where $\delta(x)$ is a general notation for a function of x which is positive and < 1.

Solution

If we take (3.21) and substitute $\alpha = 1$ and $y = x$ the expression leads to

$$\ln(1 + x) = x - x^2/2 + (x^3/3)(1 + \theta_1 x)^{-3} \quad (0 < \theta_1 < 1)$$
$$= x - x^2/2 + (x^3/3)\delta(x).$$

Putting $y = -x$ and $\alpha = 1$ on the other hand gives

$$\ln(1 - x) = -x - x^2/2 - (x^3/3)(1 + \theta_2 x)^{-3} \quad (0 < \theta_2 < 1)$$
$$= -x - x^2/2 - (x^3/3)\delta(x).$$

From the first of these expressions as n becomes large and for $x < n$

$$n \ln(1 + x/n) \rightarrow x$$

so that

$$\lim_{n \to \infty} (1 + x/n)^n = e^x \quad \text{and} \quad \lim_{n \to \infty} (1 - x/n)^n = e^{-x} \tag{3.24}$$

3.6 The normal distribution

Let us now ask the question, for what values of C, a and b is the function

$$g(x) = C \exp[-(x - a)^2/2b^2] \quad (-\infty < x < \infty)$$

a frequency function? To obtain an expression for C we have

$$C \int_{-\infty}^{\infty} \exp[-(x - a)^2/2b^2] \, dx = 1,$$

so that substituting $(x - a)/b = \omega$ in this integral gives

$$C \int_{-\infty}^{\infty} \exp(-\omega^2/2) \, d\omega = 1$$

It therefore follows from 4 in Section 3.5 that $C = [b\sqrt{(2\pi)}]^{-1}$. If the mean of the distribution is m then

$$m = E(X) = C \int_{-\infty}^{\infty} x \exp[-(x-a)^2/2b^2] \, dx$$

$$= [\sqrt{(2\pi)}]^{-1} \int_{-\infty}^{\infty} (\omega b + a) \exp(-\omega^2/2) \, d\omega$$

$$= [(\sqrt{(2\pi)})]^{-1} \left\{ b[-\exp(-\omega^2/2)]_{-\infty}^{\infty} + a \int_{-\infty}^{\infty} \exp(-\omega^2/2) \, d\omega \right\},$$

so that $m = a$. The variance of the distribution is therefore

$$\sigma^2 = E(X-a)^2 = C \int_{-\infty}^{\infty} (x-a)^2 \exp[-(x-a)^2/2b^2] \, dx$$

$$= [b^2/\sqrt{(2\pi)}] \int_{-\infty}^{\infty} \omega^2 \exp(-\omega^2/2) \, d\omega = b^2.$$

Thus when $g(x)$ is a frequency function we have

$$g(x) = [\sigma\sqrt{(2\pi)}]^{-1} \exp[-(x-m)^2/2\sigma^2]. \tag{3.25}$$

A variate with this f.f. is said to be a normal variate with mean m and standard σ. $N(m; \sigma^2)$ is an abbreviation used for such a variate. Instead of saying X is a normal variate with mean m and variance σ^2, we say X is $N(m; \sigma^2)$. When X is $N(0; 1)$, (that is, it has zero mean and unit standard deviation), it is called a standardized normal variate. It is customary to write the f.f. of this variate as $\phi(x)$ so that

$$\phi(x) = [\sqrt{(2\pi)}]^{-1} \exp(-x^2/2),$$

and to denote its distribution function by $\Phi(x)$ with

$$\Phi(x_1) = \int_{-\infty}^{x_1} \phi(x) \, dx = 1 - \Phi(-x_1).$$

Using this notation, a variate with f.f. (3.25) can be written as $\phi[(x-m)/\sigma]$ with d.f. $\Phi[(x-m)/\sigma]$. This variate is symmetric about its mean, so that $\beta_1(X)$ is equal to 0 whilst

$$m_4(X) = \int_{-\infty}^{\infty} (x-m)^4 \phi[(x-m)/\sigma] \, dx = 3\sigma^2,$$

giving $\beta_2(X) = 3$.

In view of these values of $\beta_1(X)$ and $\beta_2(X)$ those of a binomial variate given by eqns (3.13) and (3.14) and the observations made in Section 3.2 we ask the following question. Does the distribution of a binomial variate approach that of a normal variate as n increases so that

$$f(x; n, p) \cong [\sqrt{(2\pi npq)}]^{-1} \exp[-(x - np)^2/2npq]?$$

Using eqn (3.22) and writing $Y = X - np$ gives

$$f(y; n, p) = \{\sqrt{(2\pi npq)}[(1+y/np)^{np+y+1/2}(1-y/nq)^{nq-y+1/2}\}^{-1}. \quad (3.26)$$

Taking the expressions obtained in Example 3.9 we have for y less than the smaller of np and nq

$$(y+np)\ln(1+y/np) = y+y^2/2np - [(1/2) - (1+y/3np)\delta(y/np)](y/np)^3.$$

Evidently for a fixed value of y, as n increases so

$$(y + np)\ln(1 + y/np) \rightarrow y + y^2/2np.$$

In like manner

$$(-y + nq)\ln(1 - y/nq) \rightarrow -y + y^2/2nq.$$

Adding these two results together and substituting them in (3.26) yields

$$f(y; n, p) \cong \{\sqrt{(2\pi npq)}[(1 + y/np)^{1/2}(1 - y/nq)^{1/2}\}^{-1} \exp(-y^2/2npq)$$

which as n becomes larger gives

$$f(y; n, p) \cong [\sqrt{(2\pi npq)}]^{-1} \exp(-y^2/2npq) = \phi[(x - np)/\sqrt{(npq)}]. \quad (3.27)$$

Much statistical theory including quality control is concerned with data obtained from relatively small samples. For results like (3.27) to be useful we need the approximation to work for values of n which arise in practice. This remark leads to the question, how small can n be for (3.27) to be sufficiently accurate for practical needs? Two considerations arise to answer this question. The first is the value of n required for Stirling's approximation to be close to $n!$. We have seen that n can be as small as 10. The second consideration is the value of p. If p is in the region of 0.5, Table 3.2 indicates that n could also be as small as 10. As p moves away from its central value the binomial frequency function becomes asymmetric about its mean np. We accordingly find that for such values of p, n needs to increase for the approximation (3.27) to work. Generally speaking, the lowest value of n at which we can use the right-hand side of (3.27) to replace its left-hand side lies between 10 and 100. Table 3.3 shows comparisons between the values of both sides of (3.27) for $p = 0.50$, $n = 10$ and $p = 0.10$ with $n = 100$.

An important aspect of eqn (3.27) is its use to calculate total probabilities of the kind $p(X \geq x_1)$ and $p(x_1 < X \leq x_2)$. Figure 3.2 shows that

$$p(X \geq x_1) = \sum_{x=x_1}^{x=n} f(x; n, p)$$

Table 3.2   Values of $f(x; n, p)$ and its normal approximation

x	$n = 10$ $f(x; n.p)$	$p = 0.50$ Normal	x	$n = 100$ $f(x; n, p)$	$p = 0.10$ Normal
0	0.0010	0.0017	5	0.0339	0.0332
1	0.0098	0.0103	6	0.0596	0.0547
2	0.0439	0.0414	7	0.0889	0.0807
3	0.1172	0.1133	8	0.1148	0.1065
4	0.2051	0.2068	9	0.1304	0.1258
5	0.2461	0.2523	10	0.1319	0.1330
6	0.2051	0.2068	11	0.1199	0.1258
7	0.1172	0.1133	12	0.0988	0.1065
			13	0.0743	0.0807
			14	0.0513	0.0547
			15	0.0327	0.0332
			16	0.0193	0.0180

Table 3.3   Values of $p(X \geq x_1)$ and $p(X \geq x_2)$ and those of eqns (3.28) and (3.29)

p	n	x_1	x_2	$p(X \geq x_1)$	$p(X \leq x_2)$	Normal
0.50	10	8	2	0.0534	0.0534	0.0569
0.40	20	13	3	0.0210	0.0160	0.0200
0.30	40	18	6	0.0320	0.0238	0.0289
0.20	60	16	4	0.0308	0.0185	0.0259
0.10	100	16	4	0.0399	0.0237	0.0334

is the shaded part of the diagram and so, using (3.27),

$$p(X > x_1) \cong \int_{x_1 - 1/2}^{\infty} \phi[(x - m)/\sigma]\,dx = 1 - \Phi[(x_1 - m - 1/2)/\sigma]$$

$$= \Phi\{[(m + 1/2) - x_1]/\sigma\} \tag{3.28}$$

and

$$p(X \leq x_1) \cong \int_{-\infty}^{x_1 + 1/2} \phi[(x - m)/\sigma]\,dx = \Phi\{[(x_1 + 1/2) - m]/\sigma\}; \tag{3.29}$$

similarly,

$$p(x_2 \leq X \leq x_1) \cong \Phi\{[(x_1 + 1/2) - m]/\sigma\} - \Phi\{[(x_2 - 1/2) - m]/\sigma\} \tag{3.30}$$

where $m = np$ and $\sigma = \sqrt{(npq)}$.

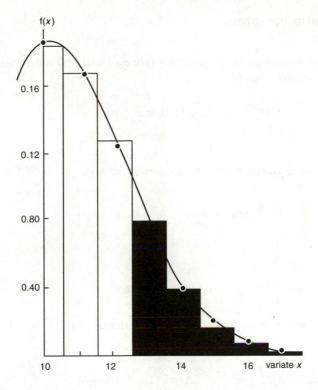

Fig. 3.2   Binomial f.f. for $n = 20$, $p = 1/2$ and the f.f. of variate $N(10; 5)$.

Table 3.4   Values of $p(X \geq x_1) + p(X \geq x_2)$ and those given by eqn (3.30)

p	n	x_1	x_2	$p(X \leq x_2)$ $+p(X \geq x_1)$	Normal
0.50	10	10	0	0.0020	0.0022
0.40	20	15	1	0.0021	0.0030
0.30	40	21	3	0.0030	0.0033
0.20	60	22	2	0.0022	0.0022
0.10	100	19	1	0.0049	0.0046

In the design of control charts we are particularly interested in these approximations when x_1 and x_2 are two or three standard deviations away from specified values of m, since $1 - \Phi(1.96) = 0.025$ and $1 - \Phi(3.09) = 0.001$. Comparisons between the left- and right-hand sides of eqn (3.29) are shown in Table 3.3 for a range of values of n and p.

The values of x_1 and x_2 in the table are the integers closest to $m \pm 2\sigma$, those for Table 3.4 correspond to $m \pm 3\sigma$.

3.7 Gamma variates

A distribution we require for the control of process variability is that of a generalized gamma variate which has f.f.

$$f(x; \lambda, \alpha) = Cx^{\lambda-1} \exp(-x/\alpha) \quad 0 < x \leq \infty$$
$$= 0 \qquad\qquad\qquad \text{otherwise}, \qquad\qquad (3.31)$$

with both λ and α greater than 0. It is customary to write

$$\Gamma(\lambda) = \int_0^\infty x^{\lambda-1} \exp(-x)\, dx \qquad\qquad (3.32)$$

so that

$$f(x; \lambda, \alpha) = [x^{\lambda-1} \exp(-x/\alpha)]/[\alpha^\lambda \Gamma(\lambda)]. \qquad\qquad (3.33)$$

For $\lambda > 1$ it follows from (3.32) that since e^{-x} dominates $x^{\lambda-1}$ as x becomes large and therefore

$$\Gamma(\lambda) = (\lambda - 1) \int_0^\infty x^{\lambda-2} e^{-x}\, dx = (\lambda - 1)\Gamma(\lambda - 1). \qquad\qquad (3.34)$$

Thus, if λ is an integer $\Gamma(\lambda) = (\lambda - 1)!$. The mean of the variate with frequency function (3.31) is

$$E(X) = \int_0^\infty \{[x^\lambda \exp(-x/\alpha)]/[\alpha^\lambda \Gamma(\lambda)]\}\, dx = \alpha\lambda$$

and its variance $V(X) = \alpha^2 \lambda$. We also have

$$m_3(X) = 2\alpha^3 \lambda \quad \text{and} \quad m_4(X) = 3\alpha^4 \lambda(\lambda + 2),$$

so that $\beta_1(X) = 2/\sqrt{\lambda}$ and $\beta_2(X) = 3 + (6/\lambda)$, and as λ increases in value $\beta_1(X) \to 0$ and $\beta_2(X) \to 3$.

If X is a $N(m; \sigma^2)$ variate then $z = [(x - m)/\sigma]^2$ is a gamma variate with parameters $\lambda = 1/2$ and $\alpha = 2$. To establish this result we note two things. First, there is a two to one relationship between x and z. Secondly, that the probability that x takes these two values is equal to the probability that z takes its corresponding value so that

$$f(z)\, dz = 2f(x)\, dx.$$

and

$$f(z)\,dz = [\sigma^{-1}\sqrt{(2/\pi)}]\exp[-(x-m)^2/2\sigma^2]\,dx$$
$$= [1/\sqrt{(2\pi)}]z^{-1/2}\exp(-z/2)\,dz.$$

We have seen that

$$\int_0^\infty \exp(-x^2/2)\,dx = \sqrt{(\pi/2)}$$

substituting $\omega = x^2/2$ in this integral gives

$$\int_0^\infty \omega^{-1/2}\exp(-\omega)\,d\omega = \sqrt{\pi} = \Gamma(1/2)$$

so that

$$f(z) = [z^{-1/2}\exp(-z/2)]/[2^{1/2}\Gamma(1/2)]. \tag{3.35}$$

Many aspects of statistical theory are needed in the design of quality control schemes. It is often necessary to fit a theoretical distribution to sampled data in order to do so, and to use a 'goodness of fit' test to indicate whether or not the sampled data came from a particular population. The generalized gamma distribution can be used for this purpose, to illustrate how this is done let us consider one or two examples.

Example 3.10

The number of accidents suffered by individual workers was recorded in a number of factories. The data obtained is given below. It could follow a Poisson distribution. However, it is reasonable to suppose that the mean accident rate for different factories will not be the same, in which case the data is more likely to follow a negative binomial distribution. Let us analyse the data to see which of these two possibilities is most likely.

Accidents per individual x	Observed number
0	447
1	132
2	42
3	21
4	3
5 and over	2

Solution

Denote the observed frequency of $X_i = x_i$ by $f_0(x_i)$ and the theoretical frequency to be tested by $f_t(x_i)$ and calculate the sum

$$\chi^2 = n \sum_{i=1}^{i=u} [f_0(x_i) - f_t(x_i)]^2 / f_t(x_i). \tag{3.36}$$

where n is the total number of events observed, and χ^2 is the ratios

(observed number of x_i − predicted number of $x_i)^2$/predicted number of x_i

summed over the u values of x_i.

Making a number of general assumptions, it can be shown that χ^2 (chi-squared) is a gamma variate with f.f.

$$f(\chi^2; v) = [(\chi^2)^{v/2-1} \exp(-\chi^2/2)]/[\Gamma(v/2)2^{v/2}]. \tag{3.37}$$

The value of v, called the number of degrees of freedom, depends on the number of constraints introduced into the data to identify the distribution fitted to it. A constraint includes the total number of observations contained in the data, in this case 647, and the number of parameters estimated from it. If this number is c, then $v = u - c$.

To test whether the data came from a Poisson distribution we need an estimate of the distribution mean. From the data given we find the mean number of accidents per persons is $m = 301/647 = 0.465$. If on the other hand, the data follow a negative binomial distribution with parameters r and p, then from eqn (3.8) its mean is rq/p whilst its variances is rq/p^2. From the observed data an estimate of the distribution variance is 0.691 which gives 0.683 as an estimate of p when $r = 1$. Thus the predictions of each distribution are as below.

x	Poisson	Negative binomial
0	406	422
1	189	140
2	44	44
3	7	14
4	1	4
5 or more		2

For the Poisson fit we see $u = 5$ and $c = 2$, so that $v = 3$ using (3.36) we find that $\chi^2 = 53.4$. Tables of the χ^2 distribution show that the probability of such a value of χ^2 if the data came from a Poisson population with mean 0.465 would be less than 0.10×10^{-5}. For the negative binomial with the same mean and variance 0.691, we see $u = 6$ and $c = 3$, and v is again equal to 3, whilst χ^2 is 4.35. The probability of this value being exceeded on 3 degrees of freedom is 0.2262. The χ^2 test of good fit therefore indicates that the data does not arise from a Poisson distribution with equivalent mean, but that it is most likely to be a mixture of these distributions.

Example 3.11

A set of 500 measurement of lengths l of manufactured items had the following number of items in the groups shown

Article length	Number observed
1.5+ to 2.50	0
2.5+ to 3.50	2
3.5+ to 4.50	3
4.5+ to 5.50	13
5.5+ to 6.50	19
6.5+ to 7.50	38
7.5+ to 8.50	63
8.5+ to 9.50	70
9.5+ to 10.5	81
10.5+ to 11.5	80
11.5+ to 12.50	52
12.5+ to 13.50	46
13.5+ to 14.50	16
14.5+ to 15.50	8
15.5+ to 16.50	6
16.5+ to 17.50	3

These measurements are supposed to come from $N(10; 6.25)$ distribution. Let us use the χ^2 test of good fit to check whether this may be so.

Solution

For a $N(m; \sigma^2)$ variate the proportion of lengths l in the range $x_1 < l \leq x_2$ will be $\Phi[(x_2 - m)/\sigma] - \Phi[(x - m)/\sigma]$ so that from tables of the normal d.f. the proportion of lengths between 13.5 and 14.5 is $\Phi(1.8) - \Phi(1.4) = 0.0449$ and the expected number of items in this group is therefore 22.5. We thus obtain the table below. The only constraint is $n = 500$, the number of degrees of freedom is 15, we therefore have $\chi^2 = 10.173$. Tables of χ^2 show that the probability that the data came from a $N(10; 6.25)$ population is high.

x_1	$500\{\Phi[(x_1 - 10)/2.5] - \Phi[(x_1 - 9)/2.5]\}$
1.5	0.07
2.5	1.7
3.5	4.6
4.5	11.0
5.5	22.5
6.5	39.0
7.5	57.8

8.5	73.2
9.5	79.3
10.5	73.2
11.5	57.8
12.5	39.0
13.5	22.5
14.5	11.0
15.5	4.6
16.5	1.7
17.5	0.7

Finally, we illustrate a difficulty which can arise when fitting a theoretical distribution to data and using the χ^2 test as a criterion of fit. Consider the following example.

Example 3.12

A sample of 1000 items from a production process showing a particular defect was as below. Their number predicted by a Poisson distribution is given in the table.

Defects	Observed number	Expected number
0	508	509.7
1	334	343.5
2	112	115.8
3	28	26.4
4	8	4.5
5		0.6

When we obtain the value of χ^2 we shall see that although its value is not significantly high there is nonetheless an unrealistic contribution to its value at the tail end of the distribution. Apart from the loss of degrees of freedom this effect can be overcome by pooling low frequency values.

Solution

The value of χ^2 for the table on 4 degrees of freedom is 3.81; of this value 3.32 comes from its last two rows. If we pool the data for 3 or more defects then the value of χ^2 drops to 1.13.

3.8 Sum of variates

Suppose we have two independent variates X_1 and X_2 with frequency function $f_1(x)$ and $f_2(x)$, what is the frequency function $f(z)$ of $Z = X_1 + X_2$? For convenience

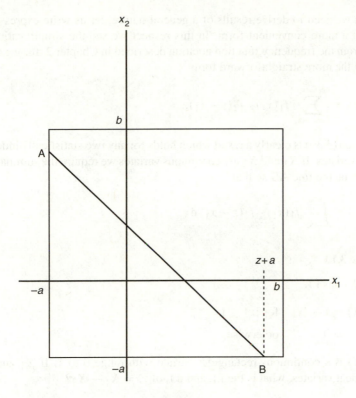

Fig. 3.3

let us assume that the variates are discrete and that both have non-zero frequencies in the range $-a < X \leq b$ where a and b are positive integers. For a point P on line AB in Fig 3.3, the probabilities of obtaining the two particular values x_1 and x_2 which are the coordinates of P are $f_1(x_1)$ and $f_2(x_2)$, so that the probability of getting the event leading to P is

$$f_1(x_1) \cdot f_2(x_2) = f_1(x_1) \cdot f_2(z - x_1).$$

The total probability of obtaining the particular sum z, is the sum of all such products on the line AB; that is, for all the integer values of x_1 between $-a$ and b, so that

$$f(z) = \sum_{x_1=-a}^{x_1=z+a} f_1(x_1) \cdot f_2(z - x_1) \quad -2a < z \leq b - a$$

and

$$f(z) = \sum_{x_1=z-b}^{x_1=b} f_1(x_1) \cdot f_2(z - x_1) \quad b - a < z \leq 2b. \tag{3.38}$$

When we need to derive results of a general nature, let us write expressions like (3.38) in a more convenient form. In this respect we see the simplification which results from the frequency function notation described in Chapter 2, for we can write (3.38) in the more straightforward form

$$f(z) = \sum_{x_1=-\infty}^{x_1=\infty} f_1(x_1) \cdot f_2(z - x_1). \tag{3.39}$$

Equation (3.39) is clearly a result which holds for any two statistically independent discrete variates. If X_1 and X_2 are continuous variates we require the summation over all points on the line AB so that

$$f(z) = \int_{-\infty}^{\infty} f_1(x_1) \cdot f_2(z - x_1) \, dx. \tag{3.40}$$

Example 3.13

Variate X_1 has f.f.

$$f(x) = 1 - |x| \quad \text{for } |x| \leq 1$$
$$= 0 \quad \quad \text{otherwise,}$$

whilst X_2 is a continuous rectangular variate with range 0 to 1. If X_1 and X_2 are independent variates, what is the f.f. and d.f. of $Z = X_1 + X_2$?

Solution

For $-1 < x_1 \leq 0$, $f_1(x_1) = 1 + x_1$ and for $0 < x_1 \leq 1$, $f_1(x_1) = 1 - x_1$ and $f_2(x_2) = 1$ for $0 < x_2 \leq 1$. Thus using (3.40) for $-1 < z \leq 0$ we have

$$f(z) = \int_{-1}^{z} (1 + x_1) \, dx_1 = z(1 + z/2) + 1/2$$

and

$$F(z_1) = \int_{-1}^{z_1} [z(1 + z/2) + 1/2] \, dz = \left\{ [z_1(1 + z_1 + z_1^2/3)]/2 \right\} + 1/6.$$

For $0 < z \leq 1$

$$f(z) = \int_{z-1}^{0} (1 + x_1) \, dx_1 + \int_{0}^{z} (1 - x_1) \, dx_1 = z(1 - z) + 1/2$$

and

$$F(z_1) = \left\{ \int_{0}^{z_1} [z(1 + z) + 1/2] \, dz \right\} + 1/6 = \{[z_1(1 + z_1 - 2z_1^3/3)/2]\} + 1/6.$$

Finally, for $1 < z \leq 2$,

$$f(z) = \int_{z-1}^{1} (1 - x_1) \, dx_1 = 2(1 - z) + z^2/2$$

and

$$F(z_1) = \left\{ \int_{1}^{z_1} [2(1 - z) + z^2] \, dz \right\} + 5/6 = z_1 \left[(2 - z_1) + z_1^2/6 \right] - 1/3.$$

3.9 Probability generating function

For discrete variate X with frequency function $f(x)$ let us introduce a dummy variable t and formulate the series

$$\alpha(t) = \sum_{x=-\infty}^{x=\infty} f(x)t^x;$$

$\alpha(t)$ is called the probability generating function of X. When $\alpha(t)$ is known, the probability that X takes a particular value x is also known, since it is the coefficient of t^x in the expansion of the series for $\alpha(t)$.

Example 3.14

What is the probability generating function of a binomial variate with parameters n and p?

Solution

We have

$$f(x)t^x = {}^nC_x(pt)^x q^{n-x}$$

so that

$$\alpha(t) = (q + pt)^n. \tag{3.41}$$

If we have two series with finite coefficients

$$\alpha_1(t) = a_0 + a_1 t + a_2 t^2 + \cdots + a_n t^n$$

and

$$\alpha_2(t) = b_0 + b_1 t + b_2 t^2 + \cdots + b_n t^n,$$

with n finite and $\alpha_1(t) = \alpha_2(t)$ for all $|t|$ less than some value $h \neq 0$, then a_i is equal to b_i for all i in the range $i = 0$ to n. It is therefore clear that the probability generating function (p.g.f.) of a discrete variate with finite range uniquely determines its frequency function. This is an example of a more general result which states if the p.g.f. of a variate X exists for all $|t| < h$ then $\alpha(t)$ uniquely determines the f.f. of X. Clearly, if two variates have the same p.g.f. they have the same f.f.

Example 3.15

Establish that

$$\alpha(t) = e^{m(t-1)}$$

is the p.g.f. of a Poisson variate with mean m

Solution

The function clearly exists over a finite range of values of t including $t = 0$, evidently

$$\alpha(t) = e^{-m} \sum_{x=0}^{x=\infty} (mt)^x / x!$$

so that applying the above theorem

$$f(x) = m^x e^{-m} / x!$$

Example 3.16

$\alpha(t)$ is the p.g.f. of a variate X which takes positive integer values only, show that

$$E(X) = \alpha'(1) \quad \text{and} \quad V(X) = \alpha''(1) + \alpha'(1)[1 - \alpha'(1)]$$

where $\alpha'(t)$ is the first differential of $\alpha(t)$ t with respect to t and $\alpha''(t)$ is its second.

A sampling procedure has one of two outcomes, success or failure. The probability of a success is p. If S is the number of samples drawn up to and including the second success, obtain the mean and variance of S by formulating its p.g.f.

Solution

We have

$$\alpha'(t) = \sum_{\text{over } x} x f(x) t^{x-1}$$

and

$$\alpha'(1) = E(X),$$

whilst

$$\alpha''(1) = \sum_{\text{over } x} x^{(2)} f(x) t^{x-2}$$

giving

$$\alpha''(1) = E(X^2) - E(X),$$

so that

$$V(X) = \alpha''(1) + \alpha'(1)[1 - \alpha'(1)].$$

For variate S we have seen that

$$f(s) = (s - 1)p^2 q^{s-2} \quad s = 2, 3, 4, \ldots, \infty$$
$$= 0 \qquad\qquad \text{otherwise.}$$

Its p.g.f. is therefore

$$\alpha(t) = (pt)^2 \sum_{s=2}^{\infty}(s - 1)(tq)^{s-2} = [(pt)^2/q]\sum_{s=2}^{\infty}(d/dt)[(tq)^{s-1}].$$

We therefore find that

$$\alpha'(t) = 2p^2 t/[(1 - qt)^3] \quad \text{and} \quad \alpha''(t) = 2p^2(1 + 2qt)/[(1 - qt)^4]$$

so that

$$\alpha'(1) = 2/p \quad \text{and} \quad \alpha''(1) = 2(1 + 2q)/p^2.$$

Using the expressions above then gives

$$E(S) = 2/p \quad \text{and} \quad V(S) = 2q/p^2.$$

3.10 The p.g.f. of a variate sum

If X_1 and X_2 are two statistically independent discrete variates, what is the p.g.f., $\alpha_3(t)$ of their sum $Z = X_1 + X_2$? From (3.39)

$$f(z)t^z = \sum_{x_1=-\infty}^{x_1=\infty} f_1(x_1)t^{x_1} f_2(z - x_1)t^{z-x_1}$$
$$= \sum_{A(z)} f_1(x_1)t^{x_1} f_2(x_2)t^{x_2}$$

where $f_1(x_1)$ and $f_2(x_2)$ are the frequency functions of X_1 and X_2 and $A(z)$ is the set of pairs x_1 and x_2 such that $x_1 + x_2 = z$, then

$$\alpha_3(t) = \sum_{\text{All } z} \sum_{A(z)} f_1(x_1)t^{x_1} f_2(x_2)t^{x_2}$$

$$= \sum_{\text{All } x_1} \sum_{\text{All } x_2} f_1(x_1)t^{x_1} f_2(x_2)t^{x_2}$$

$$= \sum_{\text{All } x_1} f_1(x_1)t^{x_1} \cdot \sum_{\text{All } x_2} f_2(x_2)t^{x_2}$$

$$= \alpha_1(t) \cdot \alpha_2(t). \tag{3.42}$$

An obvious consequence of (3.42) is that for n independent identically distributed discrete variates X_i ($i = 1$ to n) each with p.g.f. $\alpha(t)$, p.g.f. $\alpha_n(t)$ of their sum is given by

$$\alpha_n(t) = [\alpha(t)]^n. \tag{3.43}$$

The same results hold for continuous variates and can be established by replacing summations with appropriate integrals. Equation (3.43) is an extremely useful result. With its use it is a trivial matter to obtain the distribution of sums of variates whose p.g.fs. exist and are easy to formulate as the following examples illustrate.

Example 3.17

Show that the sum of n independent binomial variates each with parameters r and p is a binomial variate with parameters nr and p

Solution

From (3.41) we have seen that the p.g.f. $\alpha(t)$ of such a variate is $(q + pt)^r$. From (3.43) the p.g.f. of the sum of n of these variates $\alpha_n(t)$, is $[\alpha(t)]^n$ which is that of a binomial variate with parameters nr and p.

Example 3.18

Establish that the sum of n independent Poisson variates with means m_i; i taking values 1 to n, is a Poisson variate with mean

$$m = \sum_{i=1}^{i=n} m_i.$$

Solution

From Example 3.15, if $\alpha_i(t)$ is the p.g.f. of X_i then

$$\alpha_i(t) = \exp[m_i(t-1)]$$

and

$$\alpha_n(t) = \prod_{i=1}^{i=n} \exp[m_i(t-1)] = \exp\left[(t-1)\sum_{i=1}^{i=n} m_i\right] = \exp[(t-1)m]$$

Example 3.19

If X_1 and X_2 are two identically distributed independent negative binomial variates with parameters r and p, obtain the result that their sum Z is also a negative binomial variate with parameters 2_r and p.

Solution

The frequency function of both variates is

$$f(x) = {}^{x-1}C_{r-1}p^r q^{x-r}$$

and

$$f(x)t^x = {}^{x-1}C_{r-1}(pt)^r(qt)^{x-r},$$

so that

$$\alpha(t) = [pt/(1-qt)]^r,$$

the p.g.f. of Z is

$$[\alpha(t)]^2 = [pt/(1-qt)]^{2r}$$

whence the result follows.

Example 3.20

Each week a manufacturer receives a large shipment of steel sheets. The number of defects per sheet has a Poisson distribution whose mean m can vary from week to week. The manufacturer does not wish to reject shipments for which the mean number of defects is less or equal 0.05 more than 5 per cent of the time. He cannot afford to accept shipments whose mean number of defects is ≥ 0.20 more than 10 per cent of the time. When a shipment arrives n sheets are selected at random and are inspected. If the total number of defects exceeds r the shipment is rejected otherwise it is accepted. What values of n and r satisfy the manufacture's conditions of acceptance and rejection?

Solution

Using the result of Example 3.18, the number of defects X found in n sheets is a Poisson variate with mean nm. A similar result to eqn (3.23) for a Poisson variate with mean μ is

$$\sum_{x=0}^{x=r} \mu^x e^{-\mu}/x! = 1 - [1/\Gamma(r+1)]^{-1} \int_0^\mu \omega^r e^{-\omega} \, d\omega$$

or, in terms of the χ^2 distribution,

$$\sum_{x=0}^{x=r} \mu^x e^{-\mu}/x! = p\{[\chi^2; 2(r+1)] \geq 2\mu\};$$

that is, the probability that χ^2 on $2(r+1)$ degrees of freedom is $\geq 2\mu$.

Write $\chi_p^2[2(r+1)]$ for the value of χ^2 on $2(r+1)$ degrees of freedom for which the probability that $\chi^2[2(r+1)] \geq \chi_p^2[2(r+1)]$ is p; that is,

$$p\{\chi^2[2(r+1)] \geq \chi_p^2[2(r+1)]\} = p.$$

Clearly, this value of χ^2 is the value for which $\chi^2[2(r+1)] \leq \chi_p^2[2(r+1)]$ has probability $(1 - p)$. Thus we need the smallest values of r and n such that

$$p\{\chi^2[2(r+1)] \leq 0.10\} \leq 0.05$$

and

$$p\{\chi^2[2(r+1)] \leq 0.40\} \geq 0.90.$$

These conditions are satisfied when $\chi_p^2[2(r+1)] \geq 0.10n$ for $p = 0.05$ and when for $p = 0.90$ the value of $\chi_p^2[2(r+1)] \leq 0.40n$; that is, when

$$\{\chi_p^2[2(r+1)]_{p=0.90}\}/0.40 \leq n \leq \{\chi_p^2[2(r+1)]_{p=0.05}\}/0.10.$$

Tables of the χ^2 distribution function show that the first values of r and n which satisfy this inequality are $r = 5$ and $n = 47$.

The p.g.f. of some frequency functions is difficult to handle from an algebraic point of view. An important example is that of a normal variate. In this case a more amenable function is the moment generating function.

3.11 Moment generating function (m.g.f.)

If we replace t in $\alpha(t)$ by e^t and write the modified function as $M(t)$ so that

$$M(t) = \alpha(e^t)$$

then

$$M(t) = \sum_{-\infty}^{\infty} f(x)e^{tx} \text{ or } \int_{-\infty}^{\infty} f(x)e^{tx}dx$$

depending upon whether the variate is discrete or continuous. $M(t)$ is called the moment generating function (m.g.f.) of variate X with f.f. $f(x)$, since we can express $M(t)$ in terms of the moments of X, for example if X is discrete

$$M(t) = \sum_{x=-\infty}^{x=\infty} \sum_{r=0}^{r=\infty} f(x)(xt)^r/r! = \sum_{r=0}^{r=\infty} \mu_r(x)t^r/r!$$

and

$$[d^r/dt^r)M(t)]_{t=0} = \mu_r(x). \tag{3.44}$$

If $m(t)$ is the function which generates the mean moments of variate X with mean m, then $m(t) = e^{-mt}M(t)$. Thus if X is continuous

$$m(t) = \int_{-\infty}^{\infty} f(x)e^{t(x-m)}\,dx = e^{-mt}M(t) \tag{3.45}$$

and

$$(d^r/dt^r)m(t) = m_r(x). \tag{3.46}$$

As with the probability generating function, the moment generating function of a variate X, when it exists, determines its f.f. $f(x)$. Thus if for a positive number h, the m.g.f. of X exists for all t in the interval $-h < t < h$ then $M(t)$ uniquely determines $f(x)$. Accordingly, if two variates X_1 and X_2 have the same m.g.fs., they have the same frequency function.

If two variates X_1 and X_2 have m.g.fs., $M_1(t)$ and $M_2(t)$, the relationship which corresponds to (3.42) is that the m.g.f. $M_3(t)$ of $Z = X_1 + X_2$ is

$$M_3(t) = M_1(t) \cdot M_2(t). \tag{3.47}$$

It follows from this result that $M_n(t)$ the m.g.f. of the sum of n independent variates each with m.g.f. $M(t)$ is

$$M_n(t) = [M(t)]^n. \tag{3.48}$$

In view of the uniqueness property of the m.g.f. and this latest result we can easily determine the distribution of the sum of n independent $N(m; \sigma^2)$ variates. To obtain

$M(t)$ we have

$$M(t) = [1/\sigma\sqrt{(2\pi)}] \int_{-\infty}^{\infty} \exp[-(x-m)^2/2\sigma^2] \exp tx \, dx$$

$$= [1/\sigma\sqrt{(2\pi)}] \exp(mt + t^2\sigma^2/2) \int_{-\infty}^{\infty} \exp[-(x-m-\sigma^2t)^2/2\sigma^2] \, dx$$

$$= \exp(mt + t^2\sigma^2/2) \tag{3.49}$$

from (3.48)

$$M_n(t) = \exp[n(mt + t^2\sigma^2/2)].$$

Hence the sum of n independent identically distributed $N(m; \sigma^2)$ variates is a normal variate with mean nm and variance $n\sigma^2$. This leads to the result that the mean of n independent $N(m; \sigma^2)$ variates is an $N(m; \sigma^2/n)$ variate.

The f.f. of a generalized gamma variate X with parameters λ and α is

$$f(x) = \{1/[\Gamma(\lambda)\alpha^\lambda]\}x^{\lambda-1}\exp(-x/\alpha) \qquad x \geq 0$$

$$= 0 \qquad\qquad\qquad\qquad\qquad \text{otherwise,}$$

its m.g.f. is therefore

$$M(t) = \{1/[\Gamma(\lambda)\alpha^\lambda]\} \int_0^{\infty} x^{\lambda-1}\exp\{[x(t-(1/\alpha))]\, dx \quad (t > 1/\alpha)$$

$$= [1/(1-\alpha t)]^\lambda. \tag{3.50}$$

Thus the sum of n independent gamma variates with parameters λ and α has m.g.f. $[1/(1-\alpha t)]^{n\lambda}$ and is therefore a generalized gamma variate with parameters $n\lambda$ and α. If X_i ($i = 1$ to n) is a set of independent $N(m; \sigma^2)$ variates then its follows from this result that

$$y = \sum_{i=1}^{i=n}(x_i - m)^2$$

has f.f.

$$f(y) = \{1/[(2\sigma^2)^{n/2}\Gamma(n/2)]\}y^{(n/2)-1}\exp[-(y/2\sigma^2)]. \tag{3.51}$$

3.12 An important m.g.f. property

When we come to consider sequential schemes for the control of the parameters of a continuous variate we shall need the property which we now establish.

Take a variate X with non-zero mean, which takes both positive and negative values with non-zero frequencies. Suppose its m.g.f. $M(t)$ exists for finite t and that its first two derivatives with respect to t also exist and can be obtained by differentiation within the integral defining $M(t)$, so that

$$M''(t) = \int_{-\infty}^{\infty} (d^2/dt^2)e^{tx} f(x)\, dx.$$

For such a variate the equation $M(t) = 1$ has one non-zero root. Since X takes positive values with non-zero frequency we can find a value k such that

$$p[X \geq \ln(1+k)] > 0.$$

For t positive

$$M(t) = \int_{-\infty}^{\infty} e^{tx} f(x)\, dx = \int_{-\infty}^{\ln(1+k)} e^{tx} f(x)\, dx + \int_{\ln(1+k)}^{\infty} e^{tx} f(x)\, dx$$

$$> \int_{\ln(1+k)}^{\infty} e^{tx} f(x)\, dx > (1+k)^t\, p\{X \geq \ln(1+k)\}$$

and so

$$\lim_{t \to \infty} M(t) = \infty.$$

We can also find $k_1 > 0$ such that $p[X \leq \ln(1 - k_1)] > 0$, so that for negative t,

$$M(t) = \int_{-\infty}^{\ln(1-k_1)} e^{tx} f(x)\, dx + \int_{\ln(1-k_1)}^{\infty} e^{tx} f(x)\, dx$$

$$> \int e^{tx} f(x)\, dx > (1 - k_1)^t\, p[X \leq \ln(1 - k_1)]$$

and so

$$\lim_{t \to -\infty} M(T) = \infty.$$

These two limits and the fact that $M''(t) > 0$ and $E(X) \neq 0$ imply that there is a real non-zero value of t say t_1, such that $M(t_1) = 1$. Thus if t_m is the value t at which $M(t)$ takes its minimum value that is, (when $M'(t_m) = 0$ and $M'(0)$ is not zero), it follows that $M(t_m) < M(0) = 1$. $M(t)$ therefore decreases steadily (monotonically) between $-\infty < t \leq t_m$ and increases thereafter.

A variate which has the above features is a $N(m; \sigma^2)$ variate with $m \neq 0$. We have seen that its m.g.f. is

$$M(t) = \exp[mt + (t^2\sigma^2/2)]$$

so that $t_1 = -2m/\sigma^2$.

It is not necessary for $M(t)$ to exist over a wide range of finite t for the result to hold. Thus a f.f. we need for the control of a process variance has the form

$$f(x) = [1/\Gamma(n)]e^{-(x+\kappa)}(x+\kappa)^{n-1} \quad \text{where } \kappa > 0 \text{ and } -\kappa < x < \infty$$

$$= 0 \qquad\qquad\qquad\qquad\qquad \text{otherwise.}$$

Variate X which has this f.f. has

$$M(t) = e^{-\kappa t}(1-t)^{-n}.$$

Clearly, $M(t)$ exits for positive $t < 1$ and tends to infinity as $t \to -\infty$ and 1, so that all of the above conditions are satisfied and $t_1 = [1 - (\kappa/n)]^{-1}$.

3.13 Characteristic function

Some of the frequency functions we use in practice do not have moment generating functions because their moments of all orders do not exist. The utility of the m.g.f. when it exists leads us to search for a similar function which does exist for all f.fs. This can be formulated with a slight modification to the definition of the m.g.f. Suppose, for example, that for a continuous variate instead of

$$M(t) = \int_{-\infty}^{\infty} e^{tx} f(x)\, dx$$

we consider

$$\psi(t) = M(\iota t) = \int_{-\infty}^{\infty} e^{\iota t x} f(x)\, dx.$$

If we recall that $|e^{\iota t x}| = 1$, it is evident that $\psi(t)$ always exists whatever the form of $f(x)$, since

$$|\psi(t)| \le \int_{-\infty}^{\infty} |e^{\iota t x}| f(x)\, dx = 1.$$

$\psi(t)$ is called the characteristic function (c.f.) of $f(x)$. We find that if Z is the sum of two independent variates X_1 and X_2 with c.fs. $\psi_1(t)$ and $\psi_2(t)$ then the c.f. of $Z\psi_3(t)$ is

$$\psi_3(t) = \psi_1(t) \cdot \psi_2(t)$$

so that the c.f. $\psi_n(t)$ of the sum of n independent variates each with c.f. $\psi(t)$ is

$$\psi_n(t) = [\psi(t)]^n. \tag{3.52}$$

As with the m.g.f., the characteristic function of a variate defines its f.f. We shall not consider this function in any detail since its study involves functions of a complex variable and is outside the purpose of this monograph. We can, however, illustrate an important result with its use, namely that the central limit theorem does not hold all f.fs. Consider

$$f(x) = \{1/[\pi(1+x^2)]\} \quad -\infty < x < \infty$$

which is the distribution of the ratio of two independent standardized normal variates. This distribution, called the Cauchy distribution, has infinite moments so that its m.g.f. does not exist. We can show that its c.f. is $\psi(t) = e^{-|t|}$. If n variates X_i ($i = 1$ to n). Using eqn (3.52) we find therefore that the mean of n independent such variates is itself a Cauchy variate.

4
Effective use of sampled data

A properly designed control scheme requires use of sampled data in relation to the population being controlled. We now discuss this particular aspect of the use of sampled information in some detail. Suppose we have a sample of size n with values $X_i (i = 1 \text{ to } n)$, we formulate a function $S(X_1, X_2, X_3 \ldots X_n)$ of these values to estimate the parameter being controlled. This function is called a statistic. There are many forms that $S(X_1, X_2, X_3, \ldots, X_n)$ can take. For example, we might want to control the mean of a continuously operating manufacturing provess. $S(X_1, X_2, X_3, \ldots, X_n)$ could be the sample mean so that

$$S(X_1, X_2, X_3, \ldots, X_n) = \sum_{i=1}^{i=n} X_i/n,$$

or we could use the alternative

$$S(X_1, X_2, X_3, \ldots, X_n) = \sum_{i=1}^{i=n} w_i X_i/n$$

with w_i being a set of weights which sum to 1. On the other hand, another possibility is $(X_1 + X_s)/2$, where X_1 is the largest value of X in the sample and X_s is the smallest value. If we want to control the standard deviation σ of a process parameter with known mean m, we might use

$$S(X_1, X_2, X_3, \ldots, X_n) = \left[\sum_{i=1}^{i=n} (X_i - m)^2/n \right]^{1/2}, \qquad (4.1)$$

or $(X_1 - X_s)K$, where K is an appropriate constant such that $E[(X_1 - X_s)K] = \sigma$.

How are we to make a choice between such alternative statistics? Are some of them better than others from a control point of view? To answer this question we obviously need to define what we mean by better or best statistics. To do so we must ascertain those properties which acceptable estimates of a population parameter should possess. In order to do this let us now define a few concepts.

Notation

An economic notation is to use $\mathbf{X}$ and $g(\mathbf{X})$ for $g(X_1, X_2, X_3, \ldots, X_n)$. In this notation we can wirte eqn (4.1) as

$$S(\mathbf{X}) = \left[\sum_{i=1}^{i=n} (X_i - m)^2 / n \right]^{1/2}.$$

Using this notation let us define two basic properties which we would expect a good estimator to satisfy or nearly satisfy.

Unbiasedness

A statistic $S(\mathbf{X})$ of n variate values $\mathbf{X}$ is an unbiased estimator of a population parameter θ if its expected value is θ. Thus $S(\mathbf{X})$ is an unbiased estimator of θ if

$$E[S(\mathbf{X})] = \theta;$$

it is a biased estimator if

$$E[S(\mathbf{X})] \neq \theta.$$

We can illustrate that an infinite number of unbiased estimators can exist for a parameter by taking the first two estimators of m already given. If $\mathbf{X}$ is a sample of n independent values taken from a population with mean m write

$$S_1(\mathbf{X}) = \sum_{i=1}^{i=n} X_i / n$$

then

$$E[S_1(\mathbf{X})] = E\left[\sum_{i=1}^{i=n} X_i / n \right] = (1/n) \sum_{i=1}^{i=n} E(X_i) = m.$$

Further, if

$$S_2(\mathbf{X}) = \sum_{i=1}^{i=n} w_i X_i \quad \text{with } w_i > 0 \text{ and } \sum_{i=1}^{i=n} w_i = 1$$

then

$$E[S_2(\mathbf{X})] = E\left[\sum_{i=1}^{i=n} w_i X_i \right] = \sum_{i=1}^{i=n} w_i E(X_i) = m \sum_{i=1}^{i=n} w_i = m.$$

Since the weights w_i can take an infinite number of values there clearly exist an infinite set of unbiased estimators of m. Which of these should we choose?

Suppose the value of the population mean m is known but its variance is not. Since $E(x - m)^2 = \sigma^2$ it is reasonable to anticipate that an estimate of σ^2 from a random sample of n values $\mathbf{X}$ would be

$$S(\mathbf{X}) = \sum_{i=1}^{i=n} (X_i - m)^2/n, \qquad\qquad (4.2)$$

so that

$$E[S(\mathbf{X})] = E\left[\sum_{i=1}^{i=n} (X_i - m)^2/n\right] = \sum_{i=1}^{i=n} E(X_i - m)^2/n = \sigma^2.$$

Evidently the statistic defined by (4.2) is an unbiased estimate of σ^2. Let us now suppose that the value of the mean of the population is not known. We have seen that

$$\bar{X} = \sum_{i=1}^{i=n} X_i/n$$

is an unbiased estimate of m, so that we could substitute $\bar{X}$ for m and consider

$$S(\mathbf{X}) = \sum_{i=1}^{i=n} (X_i - \bar{X})^2/n = \left[\sum_{i=1}^{i=n} X_i^2/n\right] - \bar{X}^2.$$

Then

$$E[S(\mathbf{X})] = E\left[\sum_{i=1}^{i=n} X_i^2/n\right] - E(\bar{X}^2) = \left[\sum_{i=1}^{i=n} E(X_i^2)/n\right] - E(\bar{X}^2)$$

and since

$$E(X_i^2) = \sigma^2 + m^2 \quad \text{and} \quad E(\bar{X}^2) = (\sigma^2/n) + m^2$$

it follows that

$$E[S(\mathbf{X})] = [(n-1)/n]\sigma^2.$$

Thus

$$E\{[n/(n-1)]S(\mathbf{X})\} = \sigma^2,$$

and

$$\sum_{i=1}^{i=n} (X_i - \bar{X})^2/(n-1)$$

is an unbiased estimator of σ^2.

The second property that an estimator $S(\mathbf{X})$ of θ should have is consistency. It is sensible to expect that if a set of n observations $\mathbf{X}$ is used to obtain information about a population being sampled, then we should learn more about it as the number of observations n increases. The concept of consistency is a formal expression of this observation.

Consistent estimators

A particular function $S(\mathbf{X})$ of observations $\mathbf{X}$ is a consistent estimator of population parameter θ if, as n increases, the probability that $S(\mathbf{X})$ differs from θ by an arbitrary small quantity δ can be made as small as we please. Expressed formally we have

$$p[(|S(\mathbf{X}) - \theta| \geq \delta] \to 0$$

as n increases. Consistency is clearly based upon Tchebycheff's theorem described in Chapter 2. Such an estimator converges in probability to θ. It is a simple matter to establish a test to see whether an unbiased estimator $S(\mathbf{X})$ of θ is consistent. If $V[S(\mathbf{X})]$ the variance of $S(\mathbf{X})$ tends to zero as n increases then $S(\mathbf{X})$ is consistent. Using Tchebycheff's inequality, namely eqn (2.17),

$$p[(|S(\mathbf{X}) - \theta| \geq k\sqrt{V[S(\mathbf{X})]} \leq 1/k^2$$

writing $\delta = k\sqrt{V[S(\mathbf{X})]}$ we obtain

$$p[(|S(\mathbf{X}) - \theta| \geq \delta] \leq V[S(\mathbf{X})]/\delta^2, \qquad (4.3)$$

so that for a fixed value of δ it is clear that any unbiased estimator of θ whose variance tends to 0 as n increases is consistent. We have already seen in Chapter 3 that if X is the number of successes in an n-fold trial from a binomial population with probability of success p, then the variance $V(p_e)$ of $p_e = X/n$ is pq/n. We therefore conclude that since $E(p_e) = p$, p_e is an unbiased consistent estimator of p. Likewise,

$$\bar{X} = \sum_{i=1}^{i=n} X_i/n$$

the sample mean of n independent observations drawn randomly from a population with mean m and standard deviation σ, is an unbiased consistent estimator of m since

$$E(\bar{X}) = m \quad \text{and} \quad V(\bar{X}) = \sigma^2/n.$$

In view of the above observations let us turn to the question of whether some functions of $S(\mathbf{X})$ of $\mathbf{X}$ are better than others. To begin to answer such a question we need a measure or measures on which to base a judgement of better or best estimators. Equation (4.3) gives a strong indication of one criterion. Thus suppose we have two statistics $S_1(\mathbf{X})$ and $S_2(\mathbf{X})$ both of which are unbiased estimators of θ. Suppose further that for any sample size $V[S_2(\mathbf{X})]$ is less than $V[S_1(\mathbf{X})]$, eqn (4.3) shows that the probability that $S_2(\mathbf{X})$ will differ from θ by δ could be less than that of $S_1(\mathbf{X})$. This implies that $S_2(\mathbf{X})$ is a better estimator of θ than $S_1(\mathbf{X})$ since the value of δ is arbitrary. We can illustrate this observation by considering a simple but informative example.

Example 4.1

Variate X has f.f.

$$f(x;\theta) = 1/\theta \quad 0 < x \leq \theta$$
$$= 0 \quad \text{otherwise.} \tag{4.4}$$

What statistic can we formulate to estimate θ? Since the mean of X is $\theta/2$ let us consider

$$S_1(\mathbf{X}) = 2\sum_{i=1}^{i=n} X_i/n = 2E(X_i) = 2\int_0^\theta (x/\theta)\,\mathrm{d}x = \theta$$

since

$$E(X^2) = \int_0^\theta (x^2/\theta)\,\mathrm{d}x = \theta^2/3 \quad \text{and} \quad V[S_1(\mathbf{X})] = \theta^2/3n$$

$S_1(\mathbf{X})$ is therefore an unbiased consistent estimator of θ. An alternative estimate of θ is $S_2(\mathbf{X}) = X_1$, where X_1 is the largest value of X in the sample values $\mathbf{X}$.

Let us examine this statistic in detail by first obtaining the f.f. of X_1. The probability that a particular value of X lies in a given interval $x, x + \mathrm{d}x$ is $f(x;\theta)\,\mathrm{d}x$, whilst $F(x_1;\theta)$ is the probability that X_1 be less than x_1. Thus the probability that in n values one will take the value x_1 whilst the others are less than x_1 is

$$n[F(x_1;\theta)]^{n-1}f(x_1;\theta)\,\mathrm{d}x_1$$

and since $F(x;\theta) = x/\theta$ when $0 < x \leq \theta$, 0 for $x \leq 0$ and 1 when $x > \theta$, it follows that if x is the largest value of X in a sample of n then its f.f. $f(x;\theta)$ is

$$f(x;\theta) = nx^{n-1}/\theta^n \quad 0 < x \leq \theta$$
$$= 0 \quad \text{otherwise.} \tag{4.5}$$

It follows from (4.5) that the expected value of X_1 is $n\theta/(n+1)$ so that an unbiased estimate of θ is $(n+1)X_1/n$. Thus if $S_3(\mathbf{X})$ is this statistic then its variance $V[S_3(\mathbf{X})] = \theta^2/[n(n+2)]$. $S_3(\mathbf{X})$ is therefore an unbiased consistent estimator of θ. It is also the case that for $n > 1$, $S_3(\mathbf{X})$ has considerably smaller variance than $S_1(\mathbf{X})$.

Example 4.2

Let us return to the example already mentioned where the population mean m is to be estimated by the weighted sum

$$S_1(\mathbf{X}) = \sum_{i=1}^{i=n} w_i X_i. \tag{4.6}$$

We have already seen that since a particular w_i can take any value between 0 and 1 subject to the constraint of their sum over n being 1, $S_1(\mathbf{X})$ defines an infinite set of statistics all of which are unbiased estimators of m. We evidently need a criterion to decide what values of w_i we should take.

We find that if we put all of them equal to one another, that is, give each X_i equal weight, then the estimator so defined,

$$S_2(\mathbf{X}) = \sum_{i=1}^{i=n} X_i/n, $$

has the smallest variance of the set (4.6). Thus if $V(X)$ is the variance of each X then $V[S_1(\mathbf{X})]$ is

$$V[S_1(\mathbf{X})] = V(X)_i \sum_{i=1}^{i=n} w_i^2. \tag{4.7}$$

Using the method of Lagrange multipliers the values of w_i which minimizes (4.7) are those which satisfy the equation

$$(\partial/\partial w_i) \left\{ V[S_1(\mathbf{X})] + \lambda \left[\left(\sum_{i=1}^{i=n} w_i \right) - 1 \right] \right\} = 0, \tag{4.8}$$

where λ is a constant and the weights w_i are subject to the constraint

$$\left(\sum_{i=1}^{i=n} w_i \right) - 1 = 0. \tag{4.9}$$

From (4.7) and (4.8) it follows that

$$w_i = -\lambda/[2V(X)]$$

and from (4.9)

$$\lambda = -2V(X)/n$$

so that $w_i = 1/n$ and $\bar{X} = S_2(\mathbf{X})$ has minimum variance and is consistent. Notice that this result does not depend upon the form of the f.f. of X so that for any f.f. of X the sample mean has minimum variance of the set of estimators defined by eqn (4.6).

These examples illustrate the existence of different estimators of a particular population parameter θ. We are therefore led to ask the question, what criterion or criteria do we need to ensure that we use sampled data effectively? We could start by measuring how close an estimator is likely to be to the true value of θ.

If X is the number of successes in a sample of n items from a binomial population, we have seen from Chapter 3 that $p_e = X/n$ is an unbiased consistent estimator of p. Statistics such as p_e and the sample mean just described are, of course, random variables. Accordingly, some sample values p_e will be closer to p than others. A criterion of closeness or proximity of p_e to p, or the sample mean to m, must therefore be based on the average value of an appropriate measure, which will indicate how close, in the long run, these statistics are to the parameters they estimate. A commonly used measure is mean squared error (m.s.e.)

This is the average value of the squared difference (distance) between the estimator and the actual parameter value. Thus the mean squared error of p_e is

$$E[(p_e - p)^2] = E\{[(X/n) - p]^2\} = V(X)/n = pq/n$$

whilst

$$S(\mathbf{X}) = \sum_{i=1}^{i=n}(X_i - \bar{X})^2/(n-1)$$

has mean squared error $E[S(\mathbf{X}) - \sigma^2]$. Notice that if we use $(\text{m.s.e.})^{1/2}$ as the measure of closeness it is in comparable units to the population parameter, for example $(\text{m.s.e. of } p_e)^{1/2} = \sigma(p_e)$. From its definition the mean squared error of an unbiased estimator is equal to its variance. In view of this equivalence and the strong possibility that we shall use unbiased estimators of a population parameter θ, what is the point of defining a measure of closeness which is then equivalent to variance? A justification is that it provides a criterion to assess the deviation of biased estimators. There are circumstances where some estimators with minimal bias have properties which override those of unbiased ones. When this is the case we can use m.s.e. as a selection criterion when a number of biased alternatives need consideration.

Let us return to unbiased estimators of a population parameter. On the basis of the probability considerations implied by Tchebycheff's theorem we should select an estimator with the smallest variance. Thus for the estimators defined by (4.6) this would lead us to choose the unweighted sample mean. With regard to our first example namely Example 4.1, we would not choose this statistic since $(n+1)X_1/2n$ has smaller variance. Notice an important aspect of the last remarks with particular reference to this example (that is, 4.1). In that example we have a set of sample values $x_1, x_2, \ldots, x_n$ where the range of values of X lies between 0 and θ. Once the value of X_1 is known all the information in the sample relevant to θ has been used. None of the remaining $(n-1)$ values of X contributes any information about its value. It is accordingly pointless to use twice the sample mean as an estimate of θ.

We can evidently formulate different functions $S(\mathbf{X})$ of $\mathbf{X}$ to estimate population parameters such as mean and variance. When obtaining the minimum variance estimator of the set (4.6) we had no need to use the form of the f.f. $f(x; \theta)$ of the variate X. An obvious question arises from this observation and Example 4.1, namely, do f.fs. $f(x; \theta)$ exist for which the unweighted sample mean is the estimator of all possible estimators of the population mean with minimal variance? Clearly, we have seen from Example 4.1 that this proposition does not hold for all frequency functions. However, there may be some particular frequency functions for which the result holds. The proposition that there are, leads to a more general question with regard to unbiased estimators $S(\mathbf{X})$ of a population parameter θ. For such a set of estimators is there a limit $L(\theta)$ such that whatever the form of $S(\mathbf{X})$

$$V[S(\mathbf{X})] \geq L(\theta)?$$

Under certain general conditions (relating to $f(x; \theta)$) the following result holds. For an unbiased estimator $S(\mathbf{X})$ of n independent values $x_1, x_2, \ldots, x_n$ with common f.f. $f(x; \theta)$

$$V[S(\mathbf{X})] \geq \{nE[(\partial/\partial\theta)\ln f(x; \theta)]^2\}^{-1}$$
$$= \{nE[(\partial^2/\partial\theta^2)\ln f(x; \theta)]\}^{-1} = L(\theta). \tag{4.10}$$

The equality sign holds in (4.10) when

$$\sum_{i=1}^{i=n}(\partial/\partial\theta)\ln f(x_i; \theta) = -\{nE[(\partial^2/\partial\theta^2)\ln f(x; \theta)]\}[S(\mathbf{X}) - \theta] \tag{4.11}$$

this equation is important when it holds since it determines the functional form of $S(\mathbf{X})$. Generally speaking, the lower bound $L(\theta)$ for $V[S(\mathbf{X})]$ exists if $f(x; \theta)$ does not have discontinuities which are functions of θ. $L(\theta)$ is called the Cramer–Rao lower bound.

4.1 Efficient estimators

A statistic $S(\mathbf{X})$ which has variance equal to $L(\theta)$ and therefore satisfies eqn (4.11) is said to be efficient. More generally the efficiency of an estimator $S(\mathbf{X})$ of θ is the ratio

$$L(\theta)/V[S(\mathbf{X})]. \tag{4.12}$$

When designing a quality control procedure to monitor θ we need to decide what function $S(\mathbf{X})$ of the sample values $x_1, x_2, \ldots, x_n$ we should use to estimate θ. When an efficient statistic exists it clearly makes sense to use it, since in terms of variance as a criterion of choice no better estimator can be found. If the lower bound cannot be achieved then we need to formulate a statistic with variance which makes the ratio (4.12) as close to 1 as possible.

Let us consider some estimates of population parameters which we shall need in the design of quality control charts. Suppose first of all we wish to control the proportion of defective items p produced by a particular plant. Just to remind ourselves we have n items taken at random from a production line, the score 1 is assigned to a defective item and 0 to a non-defective one. Thus we have n values $X_1, X_2, \ldots, X_n$ where X_i takes one of two values 0 or 1 so that

$$f(x; p) = p^x q^{1-x} \quad x = 0 \text{ or } 1$$
$$= 0 \qquad \text{otherwise.}$$

We therefore have

$$(\partial/\partial p) \ln f(x; p) = (x - p)/(pq)$$

and

$$-\{nE[(\partial^2/\partial p^2) \ln f(x; p)]\}^{-1} = pq/n$$

so that from (4.11) we conclude that $S(\mathbf{X}) = \sum_{i=1}^{i=n} X_i/n$ is an efficient estimator of p with variance pq/n.

Consider variate X which is normally distributed with f.f.

$$f(x; m, \sigma^2) = [\sigma \sqrt{(2\pi)}]^{-1} \exp[-(x - m)^2/(2\sigma^2)] \quad -\infty < x < \infty.$$

If we have a sample of n independent observations X_i taken from this population what function $S(\mathbf{X})$ should we use to estimate m? We have

$$(\partial/\partial m) \ln f(x; m, \sigma^2) = (x - m)/\sigma^2$$

whilst

$$-\{nE[(\partial^2/\partial m^2) \ln f(x; m, \sigma^2)]\}^{-1} = \sigma^2/n$$

and so eqn (4.11) gives

$$S(\mathbf{X}) = \sum_{i=1}^{i=n} X_i/n = \bar{X}$$

as an efficient estimator of m. Does an efficient estimator of the variance $v = \sigma^2$ exist when the value of m is known? If we write $Z_i = (X_i - m)^2$, then in similar fashion to the derivation of eqn (3.35),

$$f(z; v) = [\Gamma(1/2)\sqrt{(2v)}]^{-1} z^{-1/2} \exp[-z/(2v)]$$
$$(\partial/\partial v) \ln f(z; v) = [(z/v) - 1] \cdot [1/(2v)]$$

and

$$(\partial^2/\partial v^2) \ln f(z; v) = [1/(v^2)][(z/v) - 1/2]$$

since $E(Z) = v$ it follows that the lowest value of the variance of an estimator of v can have is

$$-\{nE[(\partial^2/\partial v^2) \ln f(z; v)]\}^{-1} = 2v^2/n \qquad (4.13)$$

and using eqn (4.11) we obtain that

$$S(\mathbf{X}) = \sum_{i=1}^{i=n} (X_i - m)^2/n$$

is an efficient estimator of σ^2.

If the value of m is unknown we have seen that an unbiased estimator of σ^2 is

$$\sum_{i=1}^{i=n} (X_i - \bar{X})^2/(n-1). \qquad (4.14)$$

From the remarks made in Example 3.11 concerning constraints on data and the result of eqn (3.51) it follows that since the sum

$$\sum_{i=1}^{i=n} (X_i - \bar{X})^2 \qquad (4.15)$$

is subject to the constraint

$$\sum_{i=1}^{i=n} X_i = \bar{X}$$

this statistic is distributed as χ^2 with $(n-1)$ degrees of freedom. If therefore we write Z_n for the sum (4.15) then

$$f(z_n; v) = \{(2v)^{(n-1)/2} \Gamma[(n-1)/2]\}^{-1} z_n^{(n-3)/2} \exp[-z_n/(2v)]. \qquad (4.16)$$

We know that the expected value of statistic (4.14) is v so that using this last result if we denote (4.14) by $S_1(\mathbf{X})$ then

$$V[S_1(\mathbf{X})] = 2v^2/(n-1).$$

This statistic is therefore an estimator of v which is unbiased but only asymptotically efficient; that is, it approaches efficiency as n increases. In view of this result we might be tempted to believe that $[S_1(\mathbf{X})]^{1/2}$ is an unbiased estimate of σ. However, using (4.16) we find

$$E\{[S(\mathbf{X})]^{1/2}\} = [\sigma\Gamma(n/2)\sqrt{2}]/\{\Gamma[(n-1)/2]\sqrt{(n-1)}\}$$

so that $[S(\mathbf{X})]^{1/2}$ a biased estimator of σ whilst $\alpha_n[S(\mathbf{x})]^{1/2}$ is not when

$$\alpha_n = \{\Gamma[(n-1)/2]\sqrt{(n-1)}\}/[\sqrt{2}\Gamma(n/2)].$$

Let us use a closer approximation to $\Gamma(n+1)$ than eqn (3.22) namely

$$\Gamma(n+1) \cong [\sqrt{(2\pi)}]n^{n+1/2}\,\mathrm{e}^{-n}[1 + 1/(12n)]. \tag{4.17}$$

Substitution of this expression and ignoring powers of $(1/n)$ greater than or equal to 2 leads to the conclusion that

$$S_2(\mathbf{X}) = \left[\sum_{i=1}^{i=n}(X_i - \bar{X})^2/(n - 3/2)\right]^{1/2} \tag{4.18}$$

is very nearly an unbiased estimate of σ. We find that

$$\{-nE[(\partial^2/\partial\sigma^2)\ln f(x;\sigma)]\}^{-1} = \sigma^2/(2n)$$

and

$$V[S_2(\mathbf{X})] = \sigma^2/(2n-3).$$

4.2 Hypothesis testing

To make further progress towards the design of control schemes which make efficient use of the information contained in sampled data we need to examine some aspects of testing statistical hypotheses. For simplicity we shall first of all return to the simplified situation described in Chapter 1. We consider testing two alternative propositions concerning a population parameter θ, namely that its value is either θ_0 or θ_1. We have n observations sampled randomly from a population and on the basis of this information we have to decide whether the population sampled has $\theta = \theta_0$ or $\theta = \theta_1$. We have seen that we cannot conclude with certainty that $\theta = \theta_0$ or θ_1, all we can do is obtain a probability that θ takes one of these two values. On the basis of the magnitude of these we reach a decision about the value of θ. Evidently the nearer this probability $p(\theta)$ is to 1 for one of these two values of θ, say θ_0, the stronger our belief that this was the parameter value of the population sampled. We now discuss, using some simple examples, how we can establish sensible principles to control the probabilities of drawing right and wrong conclusions.

In statistical terminology we call the possibility that $\theta = \theta_0$ (say) a statistical hypothesis. If $\theta = \theta_0$ when an industrial process is running in control or a clinical testing procedure is in control then $\theta = \theta_0$ is called the null-hypothesis. When $\theta = \theta_1$ it is called the alternative hypothesis. With regard to hypothesis testing it is useful to recognize that such testing falls into one of two categories, namely simple and composite hypotheses. In due course we shall see that the second category is particularly important to the design of 'best' control schemes.

4.3 Simple hypotheses

By definition a statistical hypothesis is a statement about the distribution of a random variable X. When this is a complete specification of the distribution of X, defining the functional form of $f(x)$ and its parameter values, we say the hypothesis is a simple hypothesis. Thus if we wish to test that

$$f(x; n, p) = {}^nC_x p^x q^{n-x}$$

with $p = 1/2$ and $n = 15$ this is a simple hypothesis, as is testing that

$$f(x; m, \sigma) = \{1/[\sigma \sqrt{(2\pi)}]\} \exp -(x - m)^2/(2\sigma^2)$$

with $m = 5$ and $\sigma = 1.5$.

It is not always easy to see how a practical problem can be formulated into a corresponding statistical hypothesis. Let us therefore take an example with regard to testing the life of a manufactured component. The first step is to convert the questions being asked into formal terms. It is currently fashionable to describe this activity as modelling. We set up a probability model which is to be tested.

Example 4.3

A production unit makes a component which is an essential part of a particular piece of equipment. If the component fails the equipment fails. Marketing experience shows that an average life of m_0 is acceptable. A new manufacturing method of producing the component becomes available from the company's research department. The cost of producing the component with the new method is much less than that currently being used. Its developers claim that the new process will certainly produce components with an average life m_0. As a matter of practical experience we know that it is usually much easier to get a new process to work under the ideal conditions of a research and development unit than those which prevail on the factory floor. Since the component is to be routinely manufactured on a mass production basis the company therefore decides to instigate a regular testing procedure to see whether the components produced by the new method still have average life m_0.

Solution

To model the problem let us assume that the probability p that a component fails in unit time is constant and not therefore a function of time t. We obviously have to obtain information about the average life of the component m by sampling a number of them. We select n items and test them to their failure time T. We have seen in Example 3.6 that under the above assumption, since $m = 1/p$ the frequency function of T, $f(t; m)$ is

$$f(t; m) = [1/m] \exp(-t/m). \tag{4.19}$$

T is therefore a generalized gamma variate with parameters 1 and m. So that from Section 3.11 the sum of n independent T values $T_1, T_2, \ldots, T_n$ is a gamma variate with parameters n and m, their mean therefore has frequency function

$$f(\bar{t}; m, n) = [(n/m)^n \bar{t}^{n-1} \exp(-n\bar{t}/m)]/ \Gamma(n) \tag{4.20}$$

and $E(\bar{T}) = m$. Checking that the average life of components is m on the above assumption is equivalent to answering the following statistical question. Is the value of the sample mean obtained from testing n components consistent with the hypothesis that the sampled population has f.f. (4.20)? When the values of n and m are specified the hypothesis becomes a simple one. Take another example:

Example 4.4

Suppose, in the treatment of a particular disease, research has led us to believe that a new drug, if produced under properly controlled conditions, should have a cure rate of 75 per cent. That is to say, that a patient treated with it will have probability of 0.75 of recovery.

Solution

If we assign a score of 1 to a patient who recovers and 0 to one who does not, we obviously convert the problem into dealing with a binomial variate which takes values 1 or 0 with probabilities 0.75 and 0.25. We can test the effectiveness of a particular batch of the drug by selecting n treated patients at random and noting the number X who recover. We ask is the value of X obtained in the sample consistent with X being a binomial variate with f.f.

$$f(x; n, 0.75) = {}^n C_x (0.75)^x (90.25)^{n-x}?$$

4.4 Composite hypotheses

In reality both hypotheses in these examples are somewhat unrealistic in the contexts from which they come. The first was concerned with the life of a manufactured component. Normally the longer a component lasts the better it is. Rather than specifying that its average life should be m_0, we would in practice wish to ensure that its minimum life should on average be m_0. Thus a more appropriate hypothesis would be that the mean of the observations $T_1, T_2, \ldots, T_n$ has f.f. (4.20) and mean $m \geq m_0$. Likewise, with Example 4.4, the greater the probability of recovery the more acceptable the treatment will be. We are really interested in whether p is greater than or equal to 0.75.

For both examples, parameters of these modified hypotheses take an infinite number of values. Hypotheses of this kind, which do not completely specify a statistical distribution, are called composite hypotheses. From a theoretical point of view it is easier to develop principles of hypothesis testing by considering simple hypotheses in the first instance. We shall accordingly begin by considering these in a little more detail.

Suppose we are concerned with a clinical situation with regard to the development of a new drug. It is either 75 per cent effective in the treatment of a particular disease or it is no better than the standard treatment which has a cure rate of 30 per cent. Of 30 patients given the drug and selected at random, 17 responded favourably. Does this result indicate that if widely used the drug will have a 75 per cent cure rate? If we compute the probability of 17 or more recoveries with $n = 30$ and $p = 0.75$ we find that

$$p(X \geq 17; 30, 0.75) = F(17; 30, 0.75) = 0.9918$$

whilst

$$p(X \geq 17; 39, 0.30) = F(17; 30, 0.30) = 0.0021.$$

From these two calculations we conclude that either we have observed a rare event and $p = 0.30$, or much more likely the sample came from a population with a rather larger value of p, namely $p = 0.75$. With this value of p repeated application of the sampling procedure would frequently result in the number of recoveries being 17 or more. In view of this remark suppose we decide to disregard events which have a very small probability of occurrence, and instead regard the outcome of a sample as yielding a result that we expect to get most of the time from the population sampled. What would be the consequences of this approach? Clearly, we shall not always reach correct conclusions. Would this be a serious matter? The answer to this question obviously depends on the consequences of occasionally being wrong. With regard to the industrial scene they will usually be economic. In medicine, however, they could be much more serious. When dealing with situations or measurements subject to chance variations. we must face the fact that erroneous conclusions are the inevitable price which has to be paid if we are to make technological progress! All we can do is relate the magnitude of the probabilities of being right or wrong to the consequences of such mistakes. It is most important in the application of quality control methods in medical research or routine testing to recognize that the procedures and probabilities used in industrial contexts are not automatically appropriate. This, in the author's experience, has been an assumption which has been made all too frequently.

To define the term significance let us consider another simple example. Suppose we have a set of 10 observations known to have been taken from a population which is $N(m; 9)$. It is thought that m, the population mean, is 5. The sampled values obtained are given in Table 4.1, do they support the hypothesis that $m = 5$?

Solution

If X_i is $N(5; 9)$ then from Section 3.11 the sample mean of 10 observations is $N(5; 0.9)$ the mean of the values in Table 4.1 is 6.205. The probability of this value or more is

$$p(\bar{X} \geq 6.205) = 1 - \Phi(1.2702) = 0.1020.$$

Thus the chance of obtaining this sample mean when $m = 5$ is 1 in 10 and is accordingly not a particularly rare occurrence. We could therefore conclude that the data in Table 4.1 does not provide evidence that the mean of the population sampled differs much from 5.

A question which arises from this example is: at what level of probability that a sample mean exceeds a particular value would we reject the value $m = 5$ as the mean of the sampled population? Suppose the results of the ten-fold sample had been those in Table 4.2 The mean of these values is 6.956 and

$$p(\bar{X} \geq 6.956) = 1 - \Phi(2.0617) = 0.0196$$

Table 4.1   Values of 10 normal variates

i	1	2	3	4	5	6	7	8	9	10
x_i	7.42	1.84	9.24	6.00	7.68	2.81	3.23	8.60	7.72	7.51

Table 4.2   Values of 10 normal values

i	1	2	3	4	5	6	7	8	9	10
x_i	6.80	7.65	9.34	5.27	0.80	9.34	10.57	5.94	10.07	5.47

so that the chance of getting a sample mean ≥ 6.956 with $m = 5$ is 1 in 51. With such a low probability of occurrence we would seriously consider rejecting the notion that $m = 5$.

4.5 Significance

Let us use α for the probability of getting a sample mean greater than a specified value, so that in the Example above the probability that the sample mean is greater than or equal 6.205 is $\alpha = 0.102$, whilst for the data of Table 4.2 it is 0.0196. At what value of α would we reject the hypothesis that $m = 5$? We could accept or reject a given hypothesis on the basis of the value of α obtained for a particular sample, or we could adopt the policy that a hypothesis is rejected only when α is less than a specified value. Which of these two procedures is used will depend upon the circumstances under which a particular hypothesis is being tested. In the design and application of quality control charts we adopt the latter approach. This is because whilst a process remains in control, testing is equivalent to successively testing the same hypothesis. In the language of statistics we might regard events as significant only if their probability of occurrence is less than a specific value, say 1 in 20. This figure is related to the central limit theorem which we have discussed in Chapter 3. It is also related to the fact that for a variate X which is $N(m; \sigma^2)$ the probability that X will deviate from m by more than 2σ is approximately 1 in 20; that is, $p(|X - m| \geq 2\sigma) \cong 0.05$. We also find that $p(|X - m| \geq 3\sigma)$ is about 1 in 500. These simple multiples of σ and the approximate probabilities associated with them lead to decision rules which are easy to formulate and use.

It is clear from our examples that as soon as we consider testing more than one hypothesis there will be a need to clarify what our objectives should be. To do so we need to define certain technical terms. Let us do this in the context of testing simple hypotheses.

Test statistic

To test which of two simple hypotheses holds in a given situation, we take a sample of observations $\mathbf{X} = (X_1, X_2, \ldots, X_n)$ and use a summarizing function of these $S(\mathbf{X})$ to decide which of the two holds. $S(\mathbf{X})$ is called the test statistic In quality control, $S(\mathbf{X})$ is an estimator of a population parameter such as the proportion of defective manufactured items, the process mean or its standard deviation. We have already seen that $S(\mathbf{X})$ can take many forms. In the design of control procedures does the use of unbiased efficient statistics produce control rules with better features than using alternative statistics $S(\mathbf{X})$? Are these criteria the ones to use in the search for tests which are about the best that can be devised to test one hypothesis against another? What do we mean by one test being better than another? We would, of course, expect efficient or asymptotically efficient statistics to be in some way involved in the design of better schemes. It will become clear that the design of such schemes splits into two parts, first the choice of statistic and second the design of a decision rule with its use. Thus when the test statistic is chosen, we ask are some rules with its use better than others ? To answer this last question we need to specify what the values of a test statistic are to be to decide when a null hypothesis holds or whether the alternative to it H_1 holds. To do this we must consider all of the values the test statistic can take and divide this set of values into two distinct parts C_1 and C_2. If $S(X)$ belongs to C_1, we accept H_0. If $S(X)$ does not belong to C_1 (that is, it belongs to C_2) we accept that H_1 is the correct hypothesis.

Critical region

We need a term to define the set of values $S(X)$ which constitute C_1 and C_2. First of all we note that C_1 and C_2 are completely separate and that C_2 is the complement of C_1, so that $C_2 = C_1{}'$. Clearly, we can drop the suffix notation and write $C_1 = C'$ and $C_2 = C$. The set of values C of the statistic $S(X)$ which result in the rejection of H_0 is called the critical region of a decision rule. Evidently the decision rule is: accept H_1 if $S(X)$ belongs to C. Such a completely defined rule defines a critical region, we shall see that, conversely, a specified critical region completely defines a decision rule.

Type I error

To illustrate one of the consequences of using the above rule, let us return to the new drug example where $p_0 = 0.30$ and H_1 is that $p = p_1 = 0.75$. If we construct the rule that when $X < 17$ accept H_0, then the critical region is all of the values of $X \geq 17$. The decision rule is accept that $p = 0.30$ if in a random sample of 30 patients less than 17 of them respond favourably to the drug. We have seen that $p(X \geq 17; 30, 0.30) = 0.0021$ so that the probability of wrongly rejecting H_0 is about 1 in 500. This probability is called a Type I error.

The number of decision rules

The definition of a critical region permits us to calculate how many different decision rules can be formulated in a given situation. When the number of values that $S(X)$ can take is finite the number of different subsets to which $S(X)$ can belong is finite. The definition of a subset to which $S(X)$ belongs constitutes a decision rule. The magnitude of the problem we face in the search for a decision rule which in some sense is better or best is indicated by the number of different rules which exist. Thus for the example we have just considered, X can take 31 values each of which can either be in a critical region or not. The total number of critical regions is therefore 2^{31}. Many of these will reject themselves as sensible rules. Even so the existence of so many alternatives emphasizes the need for formal procedures to assist in the selection of good rules.

Test power

We can obtain criteria to identify appropriate critical regions by considering the probability of rejecting a null hypothesis when alternatives to it have been formulated. A concept which is used in this context is test power. The power of a test is defined as its probability of rejecting the null hypothesis H_0. Thus the power of a test with critical region $\mathbf{C}$ is the probability that the test statistic lies in $\mathbf{C}$. Let us use $S(\mathbf{X}) \in \mathbf{C}$ to denote the statement that the statistic $S(\mathbf{X})$ lies in $\mathbf{C}$. If $\mathbf{C}$ is the critical region of a test to decide which of the two alternatives of Example 4.5 hold, and $P(\mathbf{C}; p)$ is the power of the test defined by $\mathbf{C}$ then

$$P(\mathbf{C}; p) = p[S(\mathbf{X}) \in \mathbf{C}; p] = \text{prob[test rejects } H_0; p].$$

Ideally, we would like this probability to be zero when H_0 holds and 1 when H_1 holds, so that $P(\mathbf{C}; 0.30) = 0$ and $P(\mathbf{C}; 0.75) = 1$. For the composite hypotheses $p \le p_0$ and $p \ge p_1$ we would clearly like to realize these two values for all p in the two ranges. There would be no need for the methods of statistics if these ideal conditions could be obtained. We can, however, use ideal test power as a yardstick in the identification of better control rules. To illustrate this remark let us take an example which is similar to Example 4.4.

Example 4.5

If a new drug is manufactured under properly controlled conditions it is known that it will have an 80 per cent cure rate. If these conditions are not adhered to then it is no better than the standard treatment which has a cure rate of 50 per cent. The new drug is difficult to produce involving intricate processes, it is accordingly more expensive to make and market than the latter. We wish to design a test to check the quality of a manufactured batch. Thus if p is the cure rate expressed as a proportion we have two simple hypotheses, H_0 that $p = p_0 = 0.50$ and H_1 the alternative that $p = p_1 = 0.80$. To determine which value of p a batch has, we take a random sample of patients and record the number X who respond to the drug. If $X \ge x_1$ we conclude that $p = p_1$, if $X < x_1$ we decide $p = p_0$. Thus for this rule, $\mathbf{C}$ is the region $X \ge x_1$ with x_1 taking a value between 0 and n the sample size. X is a binomial variate, so that

$$p(\mathbf{C}; p) = \sum_{x=x_1}^{x=n} {}^{n}C_x p^x q^{n-x}. \tag{4.21}$$

Take $n = 30$ and consider two critical regions $\mathbf{C_1}$ and $\mathbf{C_2}$ with $\mathbf{C_1}$ defined by $x_1 = 20$ and $\mathbf{C_2}$ by $x_1' = 21$. Equation 4.21 and tables of the binomial distribution function give $P(C_1; p_0) = 0.0494$ and $P(C_2; p_0) = 0.0214$, whilst we have $P(C_1; p_1) = 0.9744$ and $P(C_2; p_2) = 0.9389$. Which of these two critical regions should we use? A preference for one or the other will depend upon the consequences of reaching wrong conclusions. These will differ according to the context to which H_0 and H_1 apply.

Better tests

To develop concepts such as test power in relation to the design of better or best tests let us ask a fairly obvious question which arises from Example 4.5. Rather than having to accept either C_1 or C_2 is it possible to have our penny and our bun? Can we find a critical region with the best features of C_1 and C_2? Such a rule must have the low Type I error of C_2 and the power of C_1 when H_1 holds. Let us examine what happens when n is increased to 45 and C_3 is the region $X \geq 30$. Tables show that $P(C_3; p_0) = 0.0179$ and $P(C_3; p_1) = 0.9890$. This new critical region has a slightly smaller Type I error than C_2 and a small increase in the power of C_1 when $p = p_1$. If the increase in sample size from 30 to 45 can be made, C_3 evidently achieves the most desirable features of both C_1 and C_2.

Type II error

Our arithmetic indicates that some critical regions may be better than others. To make progress towards their identification we need to define a second type of error. Called a Type II error, which is the probability of rejecting H_1 when it holds. Since $P(C; p_1)$ is the probability of accepting the hypothesis that $p = p_1$ with critical region C, the Type II error of C is $1 - P(C; p_1)$. Thus the Type I error for C_3 is 0.0179 and its Type II error is 0.0110.

We shall see in Chapter 5 that when H_0 and H_1 are simple, specifying the size of these two errors leads to the determination of C and hence the corresponding decision rule.

Uniform power

There are situations, particularly in clinical testing, where the size of samples which can be taken for control purposes has an upper limit. Let us return to Example 4.5 and suppose that we are not able to increase n from 30 to 45. We may be able to increase n by 5 but that is all. Let us make H_1 more relevant to clinical testing by converting H_1 into the composite hypothesis $p > 1/2$. Suppose also that we would like to find a test with Type I error 0.02 and a Type II error 0.01. Of these two it is necessary for the Type I error to be 0.02. Examination of tables of the binomial distribution function shows that if the maximum value of n is 35 we cannot find a critical region which has these two values. What are we to do? An answer to this question can be illustrated by taking $n = 35$ and the critical region C_4 with $X \geq 24$. The features of this test and

Table 4.3 Comparison of the powers of C_2 and C_4 for $p = 1/2$ and $p > 1/2$

p	$P(C_2; p)$	$P(C_4; p)$
0.50	0.0214	0.0205
0.55	0.0694	0.0729
0.60	0.1763	0.1952
0.65	0.3575	0.4019
0.70	0.5889	0.6516
0.75	0.8034	0.8579
0.80	0.9380	0.9656

those of C_2 with $n = 30$ are shown in Table 4.3. We find that both C_2 and C_4 have almost the same Type I errors but that the power of C_4 is better than C_2 for all values of $p > 1/2$ in the table.

This table raises an important question in the context of control rules. Namely, can we formulate rules with the same Type I error but with the power of one consistently higher than the other? This question leads to a more ambitious possibility, can we find a rule with a specific Type I error with greater power than any alternative rule? Such a test, if it can be found, is said to be uniformly more powerful. The significance of these remarks in the design of efficient quality control rules is a topic we shall discuss in a later chapter.

Can we obtain conditions which will indicate the algebraic form that a statistic $S(\mathbf{X})$ should take for a decision rule based upon it to be more powerful than any alternative? To answer this question let us continue with our consideration of testing just two simple hypotheses. Suppose decisions are to be based on n independent observations $(X_1, X_2, \ldots, X_n)$ where X has f.f. $f(x; \theta)$ and $\theta = \theta_0$ is H_0 and H_1 is $\theta = \theta_1$. For a particular set of sample values $(x_1, x_2, \ldots, x_n)$ write the product

$$f(x_1; \theta) f(x_2; \theta), \ldots, f(x_n; \theta) = f(\mathbf{x}; \theta).$$

Suppose we can find a critical region $\mathbf{C}$ with the following features for $k > 0$;

(i) $f(\mathbf{x}; \theta_1) \geq k f(\mathbf{x}; \theta_0)$ for all of the samples $\mathbf{x}$ in $\mathbf{C}$;

(ii) $f(\mathbf{x}; \theta_1) < k f(\mathbf{x}; \theta_0)$ for all samples $\mathbf{x}$ which do not belong to $\mathbf{C}$;

(iii) the probability that $\mathbf{X}$ lies in $\mathbf{C}$ is α when $\theta = \theta_0$.

A critical region which meets these conditions is more powerful than any other critical region with Type I error α.

To illustrate this result let us assume that $f(x; \theta)$ is the f.f. of a continuous variate. The probability that the particular sample $\mathbf{x}$ belongs to a region $\mathbf{R}$ is the sum of all of the values $f(\mathbf{x}; \theta) \, d\mathbf{x}$ in $\mathbf{R}$. For convenience let us write

$$\int_{\mathbf{R}} f(\mathbf{x}; \theta) \, d\mathbf{x} = p(\mathbf{X} \in \mathbf{R}; \theta) = P(\mathbf{R}; \theta)$$

for this probability. If $\mathbf{C}$ is a critical region which satisfies conditions (i) to (iii), then

$$\int_{\mathbf{C}} f(\mathbf{x}; \theta_0)\, d\mathbf{x} = P(\mathbf{C}; \theta_0) = \alpha$$

and

$$\int_{\mathbf{C}} f(\mathbf{x}; \theta_1)\, d\mathbf{x} \geq k \int_{\mathbf{C}} f(\mathbf{x}; \theta_0)\, d\mathbf{x} = kP(\mathbf{C}; \theta_0).$$

Take another critical region $\mathbf{C_1}$ with Type I error α which does not entirely coincide with $\mathbf{C}$, then

$$\int_{\mathbf{C_1}} f(\mathbf{x}; \theta_0)\, d\mathbf{x} = p(\mathbf{X} \in \mathbf{C_1}; \theta_0) = P(\mathbf{C_1}; \theta_0) = \alpha.$$

Denote a region which is common to $\mathbf{C}$ and $\mathbf{C_1}$ by $\mathbf{C_{12}}$, and use $\mathbf{C} - \mathbf{C_{12}}$ for the region of $\mathbf{C}$ which is separate from $\mathbf{C_1}$. Likewise, $\mathbf{C_1} - \mathbf{C_{12}}$ is the region of $\mathbf{C_1}$, separate from $\mathbf{C}$ then

$$p(\mathbf{X} \in \mathbf{C}; \theta_1) - p(\mathbf{X} \in \mathbf{C_1}; \theta_1) = p(\mathbf{X} \in \mathbf{C} - \mathbf{C_{12}}; \theta_1) - p(\mathbf{X} \in \mathbf{C_1} - \mathbf{C_{12}}; \theta_1)$$
$$= [p(\mathbf{X} \in \mathbf{C} - \mathbf{C_{12}}; \theta_1) + kp(\mathbf{X} \in \mathbf{C_{12}}; \theta_0)]$$
$$- [p(\mathbf{X} \in \mathbf{C_1} - \mathbf{C_{12}}; \theta_1) + kp(\mathbf{X} \in \mathbf{C_{12}}; \theta_0)].$$

Using condition (i)

$$p(\mathbf{X} \in \mathbf{C} - \mathbf{C_{12}}; \theta_1) \geq kp(\mathbf{X} \in \mathbf{C_1} - \mathbf{C_{12}}; \theta_0)$$

so that

$$p(\mathbf{X} \in \mathbf{C} - \mathbf{C_{12}}; \theta_1) + kp(\mathbf{X} \in \mathbf{C_{12}}; \theta_0) > kp(\mathbf{X} \in \mathbf{C}; \theta_0) = k\alpha$$

and from condition (ii)

$$p(\mathbf{X} \in \mathbf{C_1} - \mathbf{C_{12}}; \theta_1) + kp(\mathbf{X} \in \mathbf{C_{12}}; \theta_0) < kp(\mathbf{X} \in \mathbf{C_1}; \theta_0) = k\alpha.$$

We accordingly obtain the result that $P(\mathbf{C}; \theta_1) > P(\mathbf{C_1}; \theta_1)$, and so the test defined by $\mathbf{C}$ is more powerful than the one defined by $\mathbf{C_1}$. This will evidently be the case for all alternative critical regions to $\mathbf{C}$ with the same Type I error, we therefore conclude that $\mathbf{C}$ defines a test with the highest power that can be achieved. This result is known as the Neyman–Pearson theorem.

Neyman–Pearson theorem

If for the simple hypotheses H_0 and H_1 a critical region $\mathbf{C}$ can be found such that for all points belonging to $\mathbf{C}$ and $k > 0$

$$\prod_{i=1}^{i=n} f(x_i; H_1) \geq k \prod_{i=1}^{i=n} f(x_i; H_0)$$

and for all points not belonging to $\mathbf{C}$

$$\prod_{i=1}^{i=n} f(x_i; H_1) \leq k \prod_{i=1}^{i=n} f(x_i; H_0)$$

and if the total probability that $\mathbf{X} \in \mathbf{C}$ when H_0 holds is α then $\mathbf{C}$ is a best critical region for H_0 with Type I error α.

With the help of some examples we shall see that we can use this theorem as a method to identify $S(\mathbf{X})$ and decision rules which make the most effective use of sampled data. Indeed the equation

$$\left\{ \prod_{i=1}^{i=n} f(x_i; H_1) \Big/ \prod_{i=1}^{i=n} f(x_i; H_0) \right\} \geq k$$

specifically defines $S(\mathbf{X})$ and $\mathbf{C}$ once the algebraic form of $f(x; \theta)$ is known. The theorem evidently holds for discrete variates but, of course, for obvious reasons, we may not be able to obtain a test with Type I error which is exactly equal to α.

Example 4.6

For a normal variate with known standard deviation σ consider the problem of controlling its mean at one of two levels m_0 and $m_1 (> m_0)$ using a sample of n observations. For this variate writing $f(\mathbf{x}; m)$ for the product

$$\prod_{i=1}^{i=n} f(x_i; m)$$

we have

$$[f(\mathbf{x}; m_1)/f(\mathbf{x}; m_0)] = \exp\left[-\sum_{i=1}^{i=n}(x_i - m_1)^2/2\sigma^2 \right]$$

$$\Big/ \exp\left[-\sum_{i=1}^{i=n}(x_i - m_0)^2/2\sigma^2 \right]$$

$$= \exp\{n(m_1 - m_0)[\bar{x} - (m_0 + m_1)/2]/\sigma^2\}.$$

For a critical region to be most powerful we need

$$n(m_1 - m_0)[\bar{x} - (m_0 + m_1)/2] \geq \sigma^2 \ln k$$

which gives

$$\bar{x} \geq [(m_0 + m_1)/2] + (\sigma^2 \ln k)/[n(m_1 - m_0)] = K.$$

Thus, using the Neyman–Pearson theorem we find that the most powerful test for a normally distributed variate is obtained by using the sample mean and determining the value of K which has Type I error α. We have already seen from eqn (4.12) that this statistic is an efficient estimator of m. Use of the Neyman–Pearson theorem confirms what we would intuitively anticipate, namely that an efficient estimator of a population parameter should lead to decision rules with best probability features. To find K with a specified Type I error we use the result established at the end of Chapter 3, that the sample mean of n independent identically distributed $N(m; \sigma^2)$ variates is $N(m; \sigma^2/n)$. Thus if x_α is the value of x such that $\Phi(x_\alpha) = 1 - \alpha$ then

$$(K - m_0)/(\sigma/\sqrt{n}) = x_\alpha$$

giving

$$K = m_0 + (\sigma x_\alpha/\sqrt{n}).$$

The Type II error for this critical region is

$$\int_{-\infty}^{K} \phi[(x - m_1)\sqrt{n}/\sigma]\,dx = \Phi[(K - m_1)\sqrt{n}/\sigma]. \tag{4.22}$$

To turn this example into a numerical one, suppose $m_0 = 25$, $m_1 = 40$, $\sigma = 16$, $n = 10$ and the Type I error is to be 0.05. Then $x_\alpha = x_{0.05} = 1.65$ and $K = 33.35$. Using eqn (4.22) we see that the best Type II error we can achieve is equal to $\Phi(-1.314) = 0.0944$.

Example 4.7

Let us again take X to be $N(m; \sigma^2)$. Using a sample of n observations what statistic might we use to control σ^2 when it can take one of two values, σ_0^2 and σ_1^2 with $\sigma_1 > \sigma_0$? If we write $y = (x - m)^2$ we need

$$[f(\mathbf{y}; \sigma_1^2)/f(\mathbf{y}; \sigma_0^2)] = \left\{ \exp\left[-\sum_{i=1}^{i=n} y_i/2\sigma_1^2 \right] \Big/ \exp\left[-\sum_{i=1}^{i=n} y_i/2\sigma_0^2 \right] \right\} \cdot (\sigma_0/\sigma_1)^n$$

$$= \left\{ \exp\left[(\sigma_1^2 - \sigma_0^2)\sum_{i=1}^{i=n} y_i/2(\sigma_1\sigma_0)^2 \right] \right\} \cdot (\sigma_0/\sigma_1)^n \geq k$$

thus the critical region we are looking for is

$$\bar{y} = \sum_{i=1}^{i=n} (x_i - m)^2/n \geq \{[(2n\sigma_1^2\sigma_0^2/(\sigma_1^2 - \sigma_0^2)]\ln k\} + n\ln(\sigma_1/\sigma_0)].$$

Hence the most powerful test for the control of σ^2 with m known has the form

$$\sum_{i=1}^{i=n}(X_i - m)^2/n \geq K.$$

We have seen from eqn (3.51) that this sum is a generalized gamma variate. Using this equation if the test is to have Type I error α then

$$p\left[\sum_{i=1}^{i=n}(X_i - m)^2/n \geq K\right] = [2^{n/2}\Gamma(n/2)]^{-1}\int_{nK/\sigma_0^2}^{\infty} y^{(n/2)-1}e^{-y/2}\,dy = \alpha.$$

$$(4.23)$$

If we write K_α for the value of K given by eqn (4.23), then with $n = 20$, $\alpha = 0.05$, and H_0 that $\sigma_0 = 2$, tables of χ^2 give $nK_{0.05}/\sigma_0^2 = 31.41$, so that $K_{0.05}$ is 6.28. If H_1 is that $\sigma_1 = 3.17$ we find the Type II error of the critical region defined by $K_{0.05} = 6.28$ is 0.10.

The Neyman–Pearson theorem also identifies uniformly most powerful tests when controlling H_0 to a specified Type I error whilst H_1 is a composite hypothesis. Thus suppose H_1 is that $\sigma^2 > \sigma_0^2$. Write

$$\bar{y} = K_\alpha + \delta$$

so that δ is positive when $\bar{y}$ belongs to the region $\mathbf{C}$ and is, of course, negative otherwise. For the rule $\bar{y} \geq K_\alpha$ we have

$$f(\bar{y}; \sigma)/f(\bar{y}; \sigma_0) = (\sigma_0/\sigma)^n\{\exp(\bar{n}y/2)[(\sigma^2 - \sigma_0^2)/(\sigma\sigma_0)^2]\}.$$

For a sample point on the boundary of $\mathbf{C}$, when $\delta = 0$ write

$$f(K_\alpha; \sigma)/f(K_\alpha; \sigma_0) = k(\sigma).$$

Evidently for $\sigma > \sigma_0$, $k(\sigma)$ is > 0 and we can write

$$f(\bar{y}; \sigma)/f(\bar{y}; \sigma_0) = k(\sigma)\exp\{(n\delta/2\sigma_0^2)[1 - (\sigma_0/\sigma)^2]\}$$

so that for positive δ

$$f(\bar{y}; \sigma)/f(\bar{y}; \sigma_0) \geq k(\sigma)$$

and for δ negative this ratio is $\leq k(\sigma)$, so that the region $\bar{y} \geq K_\alpha$ is a best critical region for any $\sigma^2 > \sigma_0^2$.

Example 4.8

As a further example, take X to be a discrete variate. Suppose we have a sample of n observations taken to control the proportion of defective items produced by a particular machine. Hypothesis H_0 is that $p = p_0$ whilst H_1 is that $p = p_1 > p_0$. If the number of defectives in the sample is X then

$$f(x; n, p) = {}^nC_x p^x q^{n-x}.$$

If a best critical region exists then points which belong to it will satisfy the equation

$$f(x; n, p_1)/f(x; n, p_0) = p_1^x q_1^{n-x}/(p_0^x q_0^{n-x}) \geq k \quad k > 0.$$

This ratio leads to

$$x/n \geq [\ln(q_0/q_1) + (\ln k)/n]/[\ln(p_1/p_0) + \ln(q_0/q_1)]$$

so that a best critical region exists and is the region of all points $X \geq K$ with

$$K = [n \ln(q_0/q_1) + (\ln k)]/[\ln(p_1/p_0) + \ln(q_0/q_1)]$$

and $k = p_1^K q_1^{n-K}/(p_0^K q_0^{n-K})$. If the Type I error of the test is to be α, K takes the value K_α with

$$\sum_{x=K_\alpha}^{x=n} f(x; n, p_0) = \alpha.$$

Suppose $p_0 = 0.05$ and $p_1 = 0.20$ and we want a Type I error equal to 0.10. Since X is discrete we need a value of n which gives a value of α close to 0.10. For $n = 35$ and $K = 4$ we find $\alpha = 0.0958$. The Type II error is 0.0605; for the rule reject H_0 when in a random sample of 35 items 4 or more are defective.

We can show that the test is uniformly most powerful by again writing $X = K + \delta$ then

$$f(x; n, p)/f(x; n, p_0) = k\{p(1 - p_0)/[p_0(1 - p)]\}^\delta, \tag{4.24}$$

where k takes the above value. The ratio (4.24) is evidently $> k$ when δ is positive and $< k$ when δ is negative.

We can illustrate the importance of concepts like test power and some results established in Chapter 3 in the context of quality control by considering the following example.

Example 4.9

A nurseryman markets seeds of a certain variety of plant in packets of $r = 10$ seeds. Each seed has a probability p of not germinating. He has the quality requirements that p should ideally be ≤ 0.10 and should not exceed 0.15. To check quality he decides to plant the contents of $n = 50$ packets in separate plots. If X_i is the number of seeds which do not germinate in plot i, obtain a best critical region for testing the hypothesis H_0 that $p = p_0 = 0.10$ against the alternative H_1 that $p = p_1 = 0.15$. The Type I error of the test is to be 0.05. Illustrate that the test is uniformly most powerful for $p > 0.10$.

On using the test the nurseryman finds the determination of X_i tedious and therefore decides to keep $n = 50$ but to record the number of plots having full germination. If this number falls below a given value he rejects a batch of seeds. Using a rule with Type I error 0.05, compare the power of this test procedure with that of the first test. The power of the second testing procedure can be improved by increasing the value of n. How many additional packets need to be tested for the full germination test to have similar power to the first when $p = 0.10$ and $p = 0.15$? Compare the power of this last test for the composite hypothesis that $p > p_0$.

Solution

For a best critical region with $H_0 = p_0$ and $H_1 = p_1$ we need

$$\ln[\Pi f(x_i; H_1)/\Pi f(x_i; H_0)] = \left\{\left(\sum_{i=1}^{i=n} x_i\right)\ln p_1 + \left[\sum_{i=1}^{i=n}(n-x_i)\right]\ln q_1\right\}$$
$$-\left\{\left(\sum_{i=1}^{i=n} x_i\right)\ln p_0 + \left[\sum_{i=1}^{i=n}(n-x_i)\right]\ln q_0\right\}$$
$$\geq \ln k$$

so that

$$[\ln(p_1/p_0) + \ln(q_0/q_1)]\sum_{i=1}^{i=n} x_i \geq \ln k + n\ln(q_0/q_1).$$

We therefore conclude that the region defined by $\sum_{i=1}^{i=n} X_i \geq K$ is a best critical region since K is given by

$$K = [\ln k + n\ln(q_0/q_1)]/[\ln(p_1/p_0) + \ln(q_0/q_1)]$$

and is, with this rule, greater than 0. To obtain the value of K for a specific Type I error we use the result of Example 3.17, namely that the sum of n independent binomial variates with parameters r and p is a binomial variate with parameters nr and p. Since nr is large, in this case 500, we can also use the approximation given by eqn (3.28), namely that the distribution function of a binomial variate X is closely approximated by that of a normal variate when the sample size is sufficiently large thus

$$p\left(\sum_{i=1}^{i=n} X_i \geq K\right) \cong 1 - \Phi[(K - nrp - 1/2)/(2nrpq)].$$

If the Type I error is to be 0.05 then

$$(K - nrp_0 - 1/2)/\sqrt{(nrp_0q_0)} \geq 1.645$$

with $n = 50$, $r = 10$ and $p_0 = 0.10$ this gives $K \geq 61.53$, since K has to be an integer we therefore take $K = 62$ and then

$$\Phi[(K - nrp_0 - 1/2)/\sqrt{(nrp_0q_0)}] = \Phi(1.714) = 0.9567$$

which gives a Type I error of 0.0437. When $p_1 = 0.15$ the power of the test as approximated by the normal distribution is

$$1 - \Phi[(K - nrp_1 - 1/2)/\sqrt{(nrp_1q_1)}] = 1 - \Phi(-1.6909) = 0.9545.$$

We have already seen from Example 4.8 that this test will be a uniformly most powerful one.

With regard to the second test procedure the probability P_0 of the full germination of r seeds on a plot for hypothesis H_0 is $q_0^r = 0.3486$, for H_1 the corresponding probability P_1 is 0.1969. Variate X is now the number of plots showing full germination, it is again a binomial variate but now with parameters n and P. Since n is 50 and P has the values indicated, the comparisons given in Chapter 3 indicate that we can use the normal distribution to calculate probabilities for values of X so that

$$f(x; n, P) \cong \phi[(x - nP)/\sqrt{(nPQ)}].$$

The rule which the nurseryman uses is accept that a batch of seeds has germination rate $p \geq 0.90$ only if $X \geq K$. We can obtain the value of K which has Type I error of 0.05 by using the expression

$$p(X \leq K) \cong \Phi[(K - nP + 1/2)/\sqrt{(nPQ)}]$$

and

$$K - nP + 1/2 = -1.645\sqrt{(nPQ)}$$

so that K is the integer ≤ 11.39; that is, 11. The Type I error of the test with $K = 11$, given by the normal approximation is $\Phi(-1.761) = 0.0391$ and its power for a germination rate of 0.85 is

$$\Phi[(K - nP_1 + 1/2)/\sqrt{(nP_1Q_1)}] = \Phi(0.5904) = 0.7225.$$

Evidently the power of this second test for $p_1 = 0.15$ is considerably less than the first. The power of the first test is $1 - \Phi[(K - nrp - 1/2)/\sqrt{(nrpq)}]$ for values of

p greater than p_0 with $K = 62$, that of the second is $\Phi[(K - nP + 1/2)/\sqrt{(nPQ)}]$ with $K = 11$ and $P = q^r$. Using these expressions we obtain that the power of the two tests for $p > p_0$ as follows:

p	Power of first test	Power of second test
0.10	0.0437	0.0391
0.11	0.1764	0.1059
0.12	0.4182	0.2221
0.13	0.6792	0.3817
0.14	0.8633	0.5589
0.15	0.9545	0.7225

To obtain Type I and Type II errors for the second test which are commensurate with those of the first, namely $\cong 0.05$ for $p = p_0$ and $p = p_1$, we need

$$(K - nP_0 + 1/2)/\sqrt{(nP_0Q_0)} = -1.645$$

and

$$(K - nP_1 + 1/2)/\sqrt{(nP_1Q_1)} = 1.645.$$

Solving these two equations and taking K to be an integer we find K is 23 and n is 89. The power of the test using these values of K and n is as below,

p	Power of second procedure $K = 23, n = 89$
0.10	0.0469
0.11	0.1654
0.12	0.3805
0.13	0.6337
0.14	0.8343
0.15	0.9447

Let us take another example.

Example 4.10

The quality of a particular manufactured item is assessed by measurements X of a specific feature of the article. The production process is such that the mean of X does not change, but due to the development of a machine fault its standard deviation σ can increase. It is known that X is $N(m; \sigma^2)$. Obtain a test with best power having Type I error 0.05 when $n = 24$, the in-control standard deviation σ_0 is equal to 1.00 whilst the out-of-control value is 1.50. Compute the power of this test for the composite hypothesis $\sigma_0 < \sigma \leq \sigma_1$.

Solution

We have seen from Example 4.7 that the test statistic defined by the best critical region is the sample variance

$$S^2 = \sum_{i=1}^{i=n} (X_i - m)^2/n.$$

From eqn (4.23), for a variate which is $N(m; \sigma^2)$

$$p(S^2 \geq K) = [2^{n/2}\Gamma(n/2)]^{-1} \int_{nK/\sigma^2}^{\infty} y^{(n/2)-1} e^{-y/2}\, dy.$$

Thus $K_{0.05}$, the value of K with Type I error 0.05 from tables of χ^2 is

$$nK_{0.05}/\sigma_0^2 = 36.42$$

when $n = 24$, so that $K_{0.05} = 1.52$. When $\sigma_1 = 1.5$

$$nK_{0.05}/\sigma_1^2 = 16.18$$

so that for this value of σ

$$p(S^2 \geq K_{0.05}) = [2^{12}\Gamma(12)]^{-1} \int_{16.08}^{\infty} y^{11} e^{-y/2}\, dy.$$

Tables of the gamma distribution function (for example, Biometrika Tables for Statisticians) give the value of this integral, and therefore the power of the test at $\sigma_1 = 1.5$ to be 0.8815. Values of $nK_{0.05}/\sigma^2$ for the composite hypothesis are given in Table 4.4 as are values of the test power for the corresponding values of σ.

As with the seeds man, experience gained from operating this control procedure led to the conclusion, that measuring each item is time-consuming and expensive. Accordingly a gauge was designed which would be cheaper and easy to use on the shop floor. The gauge assessed whether or not X lies in the range $m \pm d$. An item is accepted if it passes the gauge test. The rule used to decide upon the value of σ is, accept that $\sigma = \sigma_1$ if in a sample of n items r or more are rejected by the gauge. Let us examine the power of tests which use this method of controlling σ when n is kept at 24 and we retain the Type I error at 0.05.

Use p_0 for the probability that a value X will be rejected by the gauge when $\sigma = \sigma_0$ and p_1 for the corresponding probability when $\sigma = \sigma_1$. If Y is the number of items

Table 4.4 Power of the composite hypothesis

σ	1.10	1.20	1.30	1.40	1.50
$nK_{0.05}/\sigma^2$	30.10	25.29	21.55	18.58	16.18
Test power	0.1815	0.3901	0.6060	0.7740	0.8811

rejected in an n-fold sample, Y is a binomial variate with parameters n, p_0 when $\sigma = \sigma_0$, and n, p_1 when $\sigma = \sigma_1$. Let us begin by taking the value of d to be 1.15 then

$$p_0 = 2[1 - \Phi(d/\sigma_0)] = 0.25 \quad \text{and} \quad p_1 = 2[1 - \Phi(d/\sigma_1)] = 0.4433.$$

A useful method of generating the frequencies $f(y; n, p)$ is given by the ratio in Section 3.1 namely,

$$f(y; n, p) = f(y - 1; n, p)[(n - y + 1)p]/(qy). \tag{4.25}$$

Thus we find for $\sigma = \sigma_0$ that $f(0; 24, 0.25) = (0.75)^{24} = 0.0010034$ and using (4.25) $f(1; 24, 0.25) = 0.00803$. Continued use of this generating equation leads to the values of $f(y; n, p)$ given in Table 4.5.

Summing the values in Table 4.5 we find that the test with $r = 10$ has Type I error 0.0547 and its power for $\sigma_1 = 1.50$ is 0.6761.

Can we improve this power by taking a different value for d? In other words does d have an optimal value? Some exploratory calculations summarized in Table 4.6 suggest that it does. All of the tests in this table have Type I error equal to 0.0547. It shows that the best value of d is in the region of $d = 1.70$, although for all practical purposes the test power at $\sigma_1 = 1.50$ is almost the same for values of d between 1.70 and 2.00.

Table 4.5   Values of $f(y; 24, p)$ for $p_0 = 0.25$ and $p_1 = 0.4433$

y	$p_0 = 0.25$	$p = 0.4433$
0	0.00100	0.00000
1	0.00803	0.00002
2	0.03077	0.00014
3	0.07521	0.00080
4	0.13163	0.00336
5	0.17550	0.01071
6	0.18525	0.02702
7	0.15878	0.05536
8	0.11247	0.09372
9	0.06666	0.13274

Table 4.6   Power of tests for different values of d when $n = 24$ and $\sigma_0 = 1$ and $\sigma_1 = 1.5$

d	0.9154	1.150	1.445	1.706	2.094	2.435
p_0	0.3600	0.2500	0.1485	0.880	0.0363	0.0158
p_1	0.5417	0.4433	0.3354	0.2554	0.1627	0.1076
r	12	10	7	5	3	2
Power	0.5831	0.6761	0.7436	0.7709	0.7666	0.7467

Table 4.7   Power of the test $n = 33$ and $r = 6$ for the hypothesis $1 < \sigma \leq 1.5$ and $d = 1.706$

σ	1.10	1.20	1.30	1.40	1.50
Power	0.2032	0.4059	0.6137	0.7760	0.8820

Table 4.8   Values of n, r, Type I error and power at $\sigma_1 = 1.5$

d	n	r	Type I error	Power at σ_1
0.9154	55	26	0.0562	0.8773
1.150	44	16	0.0626	0.8887
1.445	35	9	0.0654	0.8787
1.706	33	6	0.0650	0.8820

We can improve the power of the 'gauge' test by increasing the value of n. Bearing in mind that n and r must be integers, we can use the normal approximation to indicate the values that n and r should take to obtain a Type I error of 0.0547 and test power of 0.8811 when $\sigma = \sigma_1 = 1.5$. We require

$$(r - np_0 - 1/2)/\sqrt{(np_0q_0)} = 1.6009$$

and

$$(r - np_1 - 1/2)/\sqrt{(np_1q_1)} = -1.1805.$$

Take $d = 1.706$ then $p_0 = 0.0880$ and $p_1 = 0.2554$. We need the values of n and r which satisfy the conditions that $(r - 0.0880n - 1/2)/0.2833\sqrt{n}$ is as close as possible to 1.6009, whilst $(r - 0.2554n - 1/2)/0.4361\sqrt{n}$ is as close to -1.1805 as it can be. For $n = 33$ and $r = 6$ the value of the first ratio is 1.5951 whilst for the second it is -1.1688. We can easily check on the accuracy of using the normal distribution in this way by utilizing the recurrence relationship (4.25). For these values of n and r we find the Type I error of the test is 0.065 and its power at $\sigma = 1.5$ is 0.8820. For the composite hypothesis that $1 < \sigma \leq 1.5$, Table 4.7 shows that the procedure using this value of d, has power which for all practical purposes is equivalent to the test relating to Table 4.4.

Our calculations indicate that provided p_0 is not too small, we can use the normal distribution to obtain values of n and r which are close to those needed to obtain a particular Type I error and power for a given value of σ_1.

Using normal approximations and subsequently eqn (4.25) we obtain Table 4.8. This gives the values of n and r of tests with more or less common Type I errors and their powers at $\sigma_1 = 1.5$ for some values of d given in Table 4.6.

These examples, and in particular Table 4.8, illustrate two important aspects of the design of rules to control quality with regard to the power which can be achieved.

They are the choice of statistic to be used and the value of parameters like d. It is clear that the proper choice of both can make a considerable difference to the power of the testing procedure and the sample size required to achieve it. There are many industrial manufacturing processes where the cost of testing is such that it matters little whether the number of items sampled is 20 odd, or 40 odd. The overriding consideration is often simplicity in the operation and interpretation of the test. Thus in the example we have just considered, counting the number of articles which pass or do not pass a gauge will usually be much more straightforward than measuring each item and calculating the sample variance. The former procedure is simple, quicker to operate and less likely to error. If sampling is cheap the need to minimize n, the sample size, does not arise, although as Table 4.8 illustrates (by an appropriate choice of d) there is no point in undertaking unnecessary testing. There are situations, particularly in clinical applications where testing is so expensive and technically difficult that the smallest possible level of sampling has to be used. When this is so the choice of statistic, appropriate values of parameters, and the method of control used is very important. We shall discuss these aspects of control chart design later; it should, however, be emphasized that simple less efficient rules also have an important role in much quality control practice.

4.6 Central limit theorem

In Chapter 3 we have seen that the f.f.

$$f(x; n, p) = {}^nC_x p^x q^{n-z} \rightarrow \phi[(x - np)/\sqrt{(npq)}]$$

as n increases. We have also seen that the sum of n independent binomial variates each with parameters r and p is a binomial variate with mean nrp and variance $nrpq$. We have therefore established the result that the sample mean of n such variates has f.f. which approaches that of a normal variate with mean rp and variance (rpq/n) as n increases.

This conclusion is only an example of a remarkable and much more general result known as the central limit theorem. This states that if a variate X has finite variance σ^2 then *whatever* the frequency function of X the sample mean of n such independent variates approaches that of a normal variate as n increases. Thus if X has mean m then

$$f(\bar{x}; m, \sigma^2) \rightarrow \{[\sqrt{(n/2\pi)}]/\sigma]\} \exp[-(\bar{x} - m)^2/(2\sigma^2/n)]$$
$$= \phi[(\bar{x} - m)/(\sigma/\sqrt{n})]$$

as n increases. This result is central to the development of statistical analysis and in particular its practical application. In many situations we conduct analyses using sample means of sets of observed measurements. If the number of observations on which each is based is sufficiently large, we can assume that the measurements being analysed are approximately normally distributed. From the comparisons made in Chapter 3 and those in this chapter, we have reason to suppose that samples do not have to be that large in practice, before we can make the normality assumption. This remark is very apt once we recognize that much of statistical analysis should be regarded as the effective use of approximations. There is an art in analysing data of the real world which can be appropriately termed the art of approximation! From the examples we have so far considered and those in the chapters which follow, we shall see that the ability to assume measurements are approximately normally distributed considerably simplifies the design of straightforward control rules.

To illustrate the manner in which the theorem is established let us consider a variate X whose moment generating function $M(t)$ exists for all of the values of t in the interval $-h < t < h$ and is continuous at $t = 0$. Suppose the mean of X is m and that its variance is σ^2. From eqn (3.45) we have

$$M(t) = e^{mt} m(t)$$

and using eqn (3.21), substituting $\alpha = 0$ gives

$$m(t) = m(0) + tm^{(1)}(0) + t^2 m^{(2)}(0)/2! + t^3 m^{(3)}(\varepsilon)/3!$$

with $\varepsilon < t$. Thus, since $m(0) = 1$, $m^{(1)}(0) = 0$, and $m^{(2)}(0) = \sigma^2$,

$$m(t) = 1 + \sigma^2 t^2/2 + t^3 m^{(3)}(\varepsilon)/6.$$

Since the m.g.f. of variate aX, where a is a constant, is $E(e^{atX}) = M(at)$ we have

$$m(t/n) = 1 + \sigma^2 t^2/(2n^2) + t^3 m^{(3)}(\varepsilon)/(6n^3),$$

where $\varepsilon < t/n$. From eqn (3.48) the m.g.f. of $\bar{X} = \sum_{i=1}^{i=n} X_i/n$ is

$$M(t/n)^n = e^{mt}[1 + \sigma^2 t^2/(2n^2) + t^3 m^{(3)}(\varepsilon)/(6n^3)]^n$$

and since $\varepsilon \to 0$ as n increases we have from eqn (3.24) of Example 3.9, that,

$$\lim_{n \to \infty.} [M(t/n)]^n = \exp[mt + \sigma^2 t^2/2n].$$

Thus, whatever the f.f of a variate X with finite mean m and variance σ^2 and m.g.f. with the above properties, the mean of n observed values approaches normality as n increases.

To get an idea of the values of n needed to assume estimates of population parameters can be taken as approximately normal, let us return to Example 4.6. Consider the case when variate X is $N(m; \sigma^2)$ and the problem of estimating its variance when m is unknown. We have seen that an unbiased estimate of σ^2 is

$$S_1(\mathbf{X}) = \sum_{i=1}^{i=n} (X_i - \bar{X})^2/(n-1).$$

The distribution of this statistic can be obtained by substituting $(n - 1)$ for n in eqn (4.23) since it can be shown that

$$\sum_{i=1}^{i=n} (X_i - \bar{X})^2 / \sigma^2$$

is a generalized gamma variate with, from eqn (3.31), $\lambda = (n - 1)/2$ and $\alpha = 2$. Using this result and substituting v for $(n - 1)$ we obtain

$$p[S_1(\mathbf{X}) \geq K] = [2^{v/2} \Gamma(v/2)]^{-1} \int_{vK/\sigma^2}^{\infty} y^{(v/2)-1} e^{-y/2} \, dy$$

and that the rth zero moment of $S_1(\mathbf{X})$ is

$$\mu_r[S_1(\mathbf{X})] = [2\sigma^2/v]^r \Gamma(r + v/2) / \Gamma(v/2).$$

Thus, as we have already seen, the expected value of $S_1(\mathbf{X})$ is α_2, whilst

$$\mu_2[S_1(\mathbf{X})] = \sigma^4(n + 1)/(n - 1),$$

giving

$$\sigma^2[S_1(\mathbf{X})] = 2\sigma^4/(n - 1).$$

Using these two results and the central limit theorem, we have that as n increases then

$$p[S_1(\mathbf{X}) \geq K] \to 1 - \Phi[(K - \sigma^2)\sqrt{(n - 1)}/(\sigma^2 \sqrt{2})]. \tag{4.24}$$

How good an approximation is the right-hand side of this equation to the probability on its left for values of n likely to be used in the control of quality? Consider both sides of this equation with regard to values of K relating to control rules Type I errors of 0.05 and 0.01 and test powers of 0.95 and 0.99 for alternative simple hypotheses. We lose no generality if we take $\sigma = 1$ then the r.h.s of (4.24) becomes

$$1 - \Phi\{(K - 1)\sqrt{[(n - 1)/2]}\}.$$

For the r.h.s. of eqn (4.24), $K_{0.05}$ is then given by $(K_{0.05} - 1)\sqrt{[(n - 1)/2]} = 1.645$. Values of $K_{0.01}$, $K_{0.05}$, $K_{0.95}$, and $K_{0.99}$ are shown in Table 4.9 for a number of values of n.

We can obtain the actual probabilities that $S_1(\mathbf{X})$ is $\geq$ to these values by using a table of the d.f. of χ^2 (*Biometrika Tables for Statisticians*). Their values are given in Table 4.10.

In view of the exact probabilities in this table and those of Table 4.9, let us examine the alternative statistic $S_2(\mathbf{X})$ of eqn (4.18) to estimate σ. We have seen that to formulate this statistic we have used the approximation

$$[(n - 1)/(n - 3/2)]^{1/2} \cong \Gamma(n/2)\Gamma[(n - 1)/2]. \tag{4.25}$$

Table 4.9   Values of $K_{0.01}$, $K_{0.05}$, $K_{0.95}$ and $K_{0.99}$

n	$K_{0.01}$	$K_{0.05}$	$K_{0.95}$	$K_{0.99}$
20	1.7560	1.5337	0.4663	0.2440
25	1.6726	1.4749	0.5251	0.3274
30	1.6119	1.4320	0.5650	0.3881
35	1.5651	1.3990	0.6010	0.4349

Table 4.10   Values of $p[S_1(\mathbf{X}) \geq K]$

n	$K_{0.01}$	$K_{0.05}$	$K_{0.95}$	$K_{0.99}$
20	0.0218	0.0638	0.9757	0.9997
25	0.0200	0.0627	0.9722	0.9992
30	0.0197	0.0619	0.9698	0.9988
35	0.0195	0.0613	0.9679	0.9933

Let us therefore begin by asking how small does n have to be for this to work? The answer is very small indeed, since if $n = 5$ the l.h.s. of (4.25) is equal to 1.3229, whilst the r.h.s. is 1.3293. We can evidently therefore take the expected value of $S_2(\mathbf{X})$ to be σ, so that since $V[S_2(\mathbf{X})] = \sigma^2/(2n - 3)$ the central limit theorem gives,

$$p[S_2(\mathbf{X}) \geq K] \cong 1 - \Phi[(K - \sigma)\sqrt{(2n - 3)}/\sigma]. \tag{4.26}$$

Furthermore, since $S_2(\mathbf{X}) = [S_1(\mathbf{X})(n - 1)/(n - 3/2)]^{1/2}$ it follows that,

$$p[S_2(\mathbf{X}) \geq K] = p\{[S_1(\mathbf{X})(n - 1)/(n - 3/2)]^{1/2} \geq K\}$$
$$= p[S_1(\mathbf{X}) \geq K^2(n - 3/2)/(n - 1)].$$

and

$$p[S_2(\mathbf{X}) \geq K] = [2^{\nu/2}\Gamma(\nu/2)]^{-1} \int_{(\nu-1/2)(K/\sigma)^2}^{\infty} y^{(\nu/2)-1}\, e^{-y/2}\, dy. \tag{4.27}$$

If we carry out calculations similar to Tables 4.9 and 4.10, $K_{0.05}$ for the normal approximation satisfies the equation $(K_{0.05} - 1)\sqrt{(2n - 3)} = 1.645$, so that for $n = 20$, $K_{0.05}$ is 1.2704. The values of K for the r.h.s. of eqn (4.26) corresponding to those of Table 4.9 are shown in Table 4.11.

Using eqn (4.27) the actual values of $p[(S_2(\mathbf{X})) \geq K]$ are contained in Table 4.12. Clearly, from this table and Table 4.10. the approximation (4.26) is considerably better than (4.24).

Table 4.11   Values of $K_{0.01}$, $K_{0.05}$, $K_{0.95}$ and $K_{0.99}$

n	$K_{0.01}$	$K_{0.05}$	$K_{0.95}$	$K_{0.99}$
20	1.3830	1.2704	0.7296	0.6170
25	1.3399	1.2402	0.7598	0.6601
30	1.3086	1.2178	0.7822	0.6914
35	1.2847	1.2010	0.7990	0.7153

Table 4.12   Values of $p[S_2(\mathbf{X}) \geq K]$

n	$K_{0.01}$	$K_{0.05}$	$K_{0.95}$	$K_{0.99}$
20	0.0126	0.0536	0.9566	0.9940
25	0.0123	0.0531	0.9559	0.9934
30	0.0122	0.0531	0.9549	0.9932
35	0.0120	0.0529	0.9545	0.9929

Table 4.13   Power comparisons using K'.

σ	1.00	1.10	1.20	1.30	1.40	1.50	1,622
K'	1.678	0.902	0.256	−0.292	−0.760	−1.167	−1.593
$1 - \Phi(K')$	0.0467	0.1835	0.3989	0.6148	0.7764	0.8786	0.9444
Power	0.0500	0.1815	0.3901	0.6060	0.7740	0.8811	0.9500

These two tables indicate that we can obtain a better normal approximation to the probability $p[S_1(\mathbf{X}) \geq K]$ than eqn (4.24), thus

$$p[S_1(\mathbf{X}) \geq K] = p\{S_2(\mathbf{X}) \geq [(n-1)K/(n-3/2)]^{1/2}\}$$

$$\cong 1 - \Phi\{\{[(n-1)K/(n-3/2)]^{1/2} - 1\}\sqrt{(2n-3)}\}.$$

For values of σ not equal to 1 we have

$$p[S_1(\mathbf{X}) \geq K] \cong 1 - \Phi\{\{[(n-1)K/(n-3/2)]^{1/2} - \sigma\}\sqrt{(2n-3)}/\sigma\}. \quad (4.28)$$

To indicate how closely this expression is likely to approximate to the power of the rule reject the null hypothesis if $S_1(\mathbf{X}) \geq K$ when n, the sample size is relatively small, let us take $n = 25$. We have seen from Example 4.10 that for Type I error equal to 0.05, $K = 1.52$, so that

$$K' = \{[(n-1)K/(n-3/2)]^{1/2} - \sigma\}\sqrt{(2n-3)}/\sigma = (1.245 - \sigma) \cdot 6.856/\sigma.$$

If we use this expression and the values given by Table 4.4 we see from Table 4.13 that an *appropriate* assumption of normality can yield close approximations to the power of a test. We shall return to this conclusion in the context of designing control charts which are straightforward to operate for the control of σ^2 or σ in the next chapter.

Part II

Principles and criteria of statistical quality control

5
Principles and criteria of statistical quality control

Both in industrial and clinical testing it is important that operatives or technicians are aware that quality or test levels are being monitored. It has been claimed that a major contributory factor in the quality of mass produced items, for example, is attributable to the appearance at unknown times of quality inspectors in white coats and with clip boards who regularly inspect control charts or VDUs installed in machine shops. Notwithstanding this remark, the value of operating properly designed control procedures has been found to be very effective indeed. Their operation, in common with the practical application of statistical methods, requires attention from the statistician managing them. Their use in real life situations frequently reveals unforeseen difficulties on the part of capable management unfamiliar with the principles on which statistical method is based. To take one of my own early experiences: I was asked to design a control procedure to check the quality of large bales of synthetic fibre which was produced to spin into a wool-fibre type of yarn. Due to the nature of the distribution of the measurements used to assess bale quality the appearance of the control chart when designed was somewhat unusual! Its visual impact was rather different to that of normal Shewhart charts of the kind we shall shortly discuss. It is possible that this may have had some influence on the minds of those responsible for using the charts to monitor bales supplied to customers. After full discussion with the appropriate specification committee an acceptable quality level emerged at which we could allow 1 in 40 bales to be wrongly rejected. In view of what subsequently happened perhaps we should remark that the rejection of a whole bale of material was an expensive decision from the point of view of the production unit. A chart was designed to meet this specification and put into operation.

The manager in charge of the unit which produced the bales of material and was accordingly responsible for operating the control procedures, was a capable intelligent individual. Before the introduction of the control charts into the plant discussions were held with him clarifying their use and the principles upon which they were based. In particular, the inevitability of wrong decisions from time to time was emphasized. A few months after the introduction of the control charts into his plant and in a state of considerable concern, the manager requested an urgent meeting of the specification committee be convened. When this took place he remarked that the method of control introduced by the statistician clearly needed modification since a serious complaint had come from a customer who had received a bale (with a particular number) which he had been unable to use. He understood that a quality control scheme had recently been introduced into the manufacturing process and was therefore surprised to receive an unusable bale of material. The control chart was produced for examination. An inspection of it showed that the bale in question was, according to the chart, quite unacceptable. As the statistician concerned I asked why the bale had been sent to the customer since using the rules of the method it should certainly have been rejected. The response to my question was interesting, informative, and not unreasonable from a non-statistician unfamiliar with statistical procedures. It was that if we looked at the chart we would see that before the bale in question 40 bales had all indicated acceptable quality levels. The one in question was the 41st and it was therefore one of the acceptable bales which would be expected to arise in about one in forty bales! This example is a good illustration of the care which needs to be exercised when applying the methods of statistics in any field.

Having indicated in Chapter 1 the kinds of control chart we can use and subsequently established the properties of various statistics that we can employ with them, let us now turn to a detailed consideration of each one. We begin with the simplest, namely a Shewhart chart with just one or two action lines. Let us do so regarding it, and the definition of acceptable and rejectable quality levels, as forming a simple model which can be used to establish principles and criteria for the design and operation of statistical control procedures and comparisons between them. It will do no harm to remind ourselves of the purpose of such charts. With regard to industrial processes Shewhart remarked that his charts could be used

- to quantify levels at which a production process might operate;
- as a device to achieve these levels;
- as a device to indicate that target levels of quality are not being met and accordingly process adjustment is required.

The fundamental purpose of statistical control is its application to processes or testing methods for which variation in measurements of quality or testing can be separated into assignable and unassignable components. Thus, suppose we take a sample of n items at regular time intervals from a given process. From the information contained in the sample we calculate a statistic X with $X = m + \varepsilon$ where ε is a random fluctuation in quality which is inherent in the process and which cannot be eradicated. If m_t is the target value for the mean of the process when it is operating satisfactorily and $X = m_t + \varepsilon$ then the process is said to be in control. If the mean of the process m changes to a value m_r which is unacceptable so that $X = m_r + \varepsilon$, the process is said to be out of control. If we can re-set the process mean to m_t after a change has been detected, the change in mean level is a rectifiable or assignable source of variation, whilst ε is clearly an unassignable source of variation. If the standard deviation of ε is σ_ε but a processing fault can cause an additional random component of error ε_1, so that variate $X = m_t + \varepsilon_1 + \varepsilon$ where ε_1 and ε are independent of one another, then $\sigma(X)$ is $(\sigma_1^2 + \sigma_\varepsilon^2)^{1/2}$. If this value of $\sigma(X)$ is not acceptable the process will be judged out of control, and if ε_1 can be removed by adjusting the production process then ε_1 is also an assignable source of variation.

Recognition of the behaviour of the component ε is, of course fundamental to the operation of statistical methods of control. If a set of values X_i $(i = 1$ to $N)$ follow the form $X = m_t + \varepsilon$ (that is, with non-systematic (random) fluctuations about m_t), the system is deemed in control when the standard deviation of these is equal to σ_ε^2. When N is large we can use the data to estimate m_t and σ_ε. Either by using these values or those obtained from experience of the process we can calculate values of X_i which will only occur rarely when a process is in control. As we have already seen we can obtain such values very simply when X_i is approximately distributed as a normal variate. We can construct a chart with vertical scale signifying values of X_i and horizontal scale denoting values of i. As described in Chapter 1, a Shewhart chart has a horizontal line showing the value of m_t and two lines, one on either side of the central line, showing the values of X_i which rarely occur when the process is in control. Suppose a significant number of values of X_i are plotted on such a chart. If all of them lie within the upper and lower lines with

- approximately the same number of results above and below the central line;
- no cyclic patterns in plotted values;
- no runs of results above or below the central line;
- no prolonged drifts of results up or down;

then we conclude that the process is in control and that the model that X_i is equal to $m_t + \varepsilon$ is valid with ε being a random component of variation. We shall indicate tests which can be used to ascertain the randomness of ε later in this chapter.

Inspection of these charts will give a visual appreciation of a process which is in control. As experience of using such charts is gained the value of visual assessment should not be underestimated. It is perhaps appropriate to emphasise this observation here, in view of the current obsession to computerise everything! We shall make further comment on this aspect of control charts later.

5.1 Control of proportion of defective items

To set the principles and criteria developed in Chapter 4 into the context of designing control charts, let us begin by considering the problem of monitoring the proportion p of defective items produced by a particular manufacturing process. Let us start with the simple model which identifies just two values of p, namely p_a as acceptable quality level and p_r as rejectable level. Entirely for ease of presentation consider a process where many thousands of items are produced in any one production run. Assume that testing is easily carried out and is cheap to perform so that the size n of a sample of say 100 items is perfectly acceptable. As in Chapter 4, we use X to denote the number of defective items in a sample of n. Recalling Example 4.8, we have seen that for a given Type I error at $p = p_a$ the most powerful test that we can find for any $p > p_a$ has the form $X \geq K$. This is precisely the rule of a Shewhart chart with only one action line, namely the upper one. To the layman, identifying values for just two values of p which can be used to formulate a most powerful test obviously makes practical sense. It presents the use of such charts in a much better light than the terms acceptable, rejectable and indifference quality levels imply. With regard to setting values for p_a and p_r we approach management by asking the following questions.

- For what values of p do you want the frequency of rejecting batches to be less than 1 in 50 (say)?
- For what range of values of p do you want the proportion of batches going out to customers to be less than 1 in 100?
- Having regard to the need to use statistically based decision rules would you accept a test based on the value of p at the end of the first range, namely p_a and that at the beginning of the second p_r, if it is the best which can be found in the sense that

  (a) it maximizes the chance of rejecting batches of material with $p > p_a$ the further p moves away from p_a;
  (b) notwithstanding (a) the chance of rejecting values of p which are marginally greater than p_a is not very different from that when $p = p_a$;
  (c) when the proportion of defective items becomes $\geq p_r$ then the chance of such material being received by customers will be less than an agreed figure?

When approached in this way we find that the identification of values for p_a and p_r is reasonably straightforward.

Example 5.1

Suppose we take Example 4.8, with the specification that the chance of rejecting batches with $p \leq p_a$ is to be 0.02 or less and that when $p \geq p_r$ is to be 0.99 or more. We therefore require the smallest value of n for which

$$\sum_{x=0}^{x=K-1} f(x; n, p) \quad \begin{cases} \geq 0.98 & \text{when } p \leq 0.05 \\ \\ \leq 0.01 & \text{when } p \geq 0.20. \end{cases} \tag{5.1}$$

Using the procedure described in Examples 4.8 and 4.9 for obtaining approximate values of n and K and subsequently refining them, we find that n is 81 and $K = 9$. The probabilities of rejecting batches for a range of values of p are given in the Table 5.1.

Table 5.1 Rejection probabilities for proportion defective Shewhart chart with $n = 81$ and $K = 9$

p	$F(8; 81, p)$
0.040	0.0050
0.050	0.0198
0.060	0.0536
0.075	0.1525
0.100	0.4220
0.125	0.6968
0.150	0.8750
0.175	0.9583
0.200	0.9896
0.220	0.9964

With regard to obtaining the value of summations like eqn (5.1) when n is large and p is small, and to emphasize the convenience as well as the importance of the central limit theorem, let us return to the result given by eqn (3.23), namely that

$$F(K - 1); n, p)$$

$$= 1 - \{\Gamma(n + 1)/[\Gamma(K)\Gamma(n - K + 1)]\} \int_0^p \omega^{K-1}(1 - \omega)^{n-K} d\omega$$

$$= \{\Gamma(n + 1)/[\Gamma(K)\Gamma(n - K + 1)]\} \int_p^1 \omega^{K-1}(1 - \omega)^{n-K} d\omega \qquad (5.2)$$

In view of this result and the content of most contemporary statistical tables produced for everyday use, consider the following example.

Example 5.2

Establish that tables of the distribution function $F(z)$ of the statistic Z can be used to evaluate the integral in eqn (5.2), where

$$F(z_1) = \{a^a b^b \Gamma(a + b)/[2^{(a+b)}\Gamma(a)\Gamma(b)]\} \int_0^{z_1} z^{a-1}/(b + az)^{a+b} \, dz. \quad (5.3)$$

Solution

If in (5.3) we substitute $y = za/b$ we have

$$F(z_1) = \Gamma(a + b)/[\Gamma(a)\Gamma(b)] \int_0^{z_1 a/b} y^{a-1}(1 + y)^{a+b} \, dy,$$

whilst if we take (5.2) and write $\omega = (1 + y)^{-1}$ we find

$$\int_p^1 \omega^{K-1}(1 - \omega)^{n-K}\,d\omega = \int_0^{q/p} y^{n-K}/(1 + y)^{n+1}\,dy$$

and

$$\sum_{x=K}^{x=n} f(x; n, p) = \Gamma(n+1)/[\Gamma(K)\Gamma(n-K+1)] \int_{q/p}^{\infty} y^{n-K}(1+y)^{n+1}\,dy. \quad (5.4)$$

The statistic Z is, in fact, the distribution of the ratio of two independent sample variances from a common population whose variate is normally distributed. Thus, if $S_1(\mathbf{X}; n_1 - 1)$ is the first of these and is based on $n_1 - 1$ degrees of freedom whilst $S_1(\mathbf{X}; n_2 - 1)$ is the second, then the distribution function of the variate defined by $Z = S_1(\mathbf{X}; n_1 - 1)/S_1(\mathbf{X}; n_2 - 1)$ is given by eqn (5.3), with $a = (n_1 - 1)/2$ and b being $(n_2 - 1)/2$. In most tables $(n_1 - 1)$ and $(n_2 - 1)$ are denoted by v_1 and v_2 so that $v_1 = 2a$ and $v_2 = 2b$.

We can use tabulated values of z_1 of the integral (5.3) for different values of v_1 and v_2 to obtain n and K for schemes with specific Type I and Type II errors. Thus suppose that we need a Type I error α when $p = p_a$ and that

$$1 - F(z_\alpha) = \alpha,$$

then from (5.4)

$$\sum_{x=k}^{x=n} f(x; n, p) = \alpha$$

when $q_a/p_a = (n - K + 1)z_\alpha/K$. For this Type I error together with a Type II error of $1 - \beta$ when $p = p_r$ we find that n and K must satisfy the equations

$$z_\alpha = (1 - p_a)K/[(n - K + 1)p_a] \qquad (5.5)$$

and

$$z_{1-\beta} = (n - K + 1)p_r/[(1 - p_r)K]. \qquad (5.6)$$

Values of z_α are given in most published Statistical Tables for values of α between 0.01 and 0.10 over a wide range of values of v_1 and v_2. Values of K and n can be obtained with their use. The form of published tables is such that in calculations involving of z_α, $v_1 = 2(n - K + 1)$ and $v_2 = 2K$, and those for $z_1 - \beta$ have $v_1 = 2K$ and $v_2 = 2(n - K + 1)$. We shall, however, find that they are not sufficiently detailed or in an appropriate form to design schemes for the values of p_a, p_r, α and $(1 - \beta)$ which arise in practice. When this is the case we can use the following expressions:

$$\ln z_\alpha \cong x_\alpha \{[1/(K - 1/2)] + [1/(n - K + 1/2)]\}^{1/2}$$
$$+ \left(x_\alpha^2 + 2\right)\{[1/(K - 1/2)] - [1/(n - K + 1/2)]\}/6 \tag{5.7}$$

$$\ln z_{1-\beta} \cong x_{1-\beta}\{[1/(K - 1/2)] + [1/(n - K + 1/2)]\}^{1/2}$$
$$- \left(x_{1-\beta}^2 + 2\right)\{[1/(K - 1/2)] - [1/(n - K + 1/2)]\}/6 \tag{5.8}$$

where $1 - \Phi(x_\alpha) = \alpha$ and $1 - \Phi(x_{1-\beta}) = 1 - \beta$. These equations are developments of those given in a paper by A. H. Carter (*Biometrika*, **34**, 352–8). For the schemes we need to design, eqns (5.7) and (5.8) have an accuracy which is sufficient to determine n and K with relative ease. If we write

$$B = [1/(K - 1/2)] + [1/(n - K + 1/2)], \tag{5.9}$$

then from eqns (5.6) to (5.8) it follows that

$$(x_\alpha + x_{1-\beta})\{\sqrt{B} + (x_\alpha - x_{1-\beta})[B - 2/(n - K + 1/2)]/6\}$$
$$\cong \ln\{(1 - p_a)p_r/[p_a(1 - p_r)]\}. \tag{5.10}$$

To indicate values of K which will give schemes we want, we can assume that the term $(x_\alpha - x_{1-\beta})/[3(n - K + 1/2)]$ is so small that it can be neglected, and then

$$(x_\alpha - x_{1-\beta})B/6 + \sqrt{B} \cong \{\ln\{(1 - p_a)p_r/[p_a(1 - p_r)]\}/(x_\alpha + x_{1-\beta})\}. \tag{5.11}$$

Values of $\sqrt{B}$ given by (5.11) for the ranges of values of n and K of our schemes are relatively invariant, and this is another reason why the method we describe below works. When we combine eqns (5.5) and (5.7) we obtain

$$x_\alpha\{[1/(K - 1/2)] + [1/(n - K + 1/2)]\}^{1/2}$$
$$+ (x_\alpha^2 + 2)\{[1/(K - 1/2)] - [1/(n - K + 1/2)]\}/6$$
$$\cong \ln\{(1 - p_a)K/[p_a(n - K + 1)]\} \tag{5.12}$$

whilst from (5.6) and (5.8)

$$x_{1-\beta}\{[1/(K - 1/2)] + [1/(n - K + 1/2)]\}^{1/2}$$
$$- (x_{1-\beta}^2 + 2)\{[1/(K - 1/2)] - [1/(n - K + 1/2)]\}/6$$
$$\cong \ln\{p_r(n - K + 1)/[(1 - p_r)K]\}. \tag{5.13}$$

Use of these equations to obtain K and n for a particular control scheme is best illustrated in the examples which follow.

Example 5.3

Suppose $p_a = 0.05$, $p_r = 0.15$, $\alpha = 0.025$ and $1 - \beta = 0.05$. From tables of the normal distribution $x_\alpha = 1.96$ and $x_{1-\beta} = 1.65$ so that (5.10) gives

$$0.05167B + \sqrt{B} \cong 0.3351$$

so that $B \cong 0.1086$ and therefore from (5.9) $K \geq 10$. If we take $K = 11$ then using 0.1086 to approximate to B eqn (5.12) gives

$$0.7516 - 1.9472/(n - 10.5) \cong \ln[209/(n - 10)].$$

whilst from eqn (5.13) we have

$$0.4582 + 1.5742/(n - 10.5) \cong \ln[(n - 10)/62.33].$$

We find that to satisfy both of these equations n has to be 110 or 111. Thus, if we take $K = 11$ and $n = 110$ we should have a scheme with Type I and Type II errors close to the values we want.

Finally, if we substitute $n = 110$ and $K = 11$ into (5.9) we have $B = 0.1053$, and substitution into eqns (5.12) and (5.13) confirms that a scheme with these values will have values of α and $1 - \beta$, very close to 0.025 and 0.05.

Let us ask the question, is this the best scheme we can find in the sense that n is as small as it can be? What is the value of n when $K = 10$? If we repeat the calculations we find that $n = 100$ and tables of the distribution function of binomial variates show that

$$\sum_{x=10}^{x=100} {}^{100}C_x p_a^x q_a^{100-x} = 0.0282 \quad \text{and} \quad \sum_{x=10}^{x=100} {}^{100}C_x p_r^x q_r^{100-x} = 0.9449.$$

Example 5.4

To take a second example, suppose $p_a = 0.10$ and $p_r = 0.20$ with $\alpha = 0.10$ and with $1 - \beta = 0.05$. Substitution into (5.11) gives

$$-0.06B + \sqrt{B} \cong 0.2758,$$

so that $B \cong 0.0787$ and from (5.9) $K \geq 13$. If we take $K = 13$ and carry through the above procedure, we find that n has to be equal to 87 to satisfy (5.12) and 93 to satisfy (5.13). On the other hand, if we take $K = 14$ the corresponding values of n are 96 and 99. Using eqn (3.7) we find that

$$\sum_{x=14}^{x=96} {}^{96}C_x p_a^x q_a^{96-x} = 0.0967 \quad \text{with} \quad \sum_{x=14}^{x=96} {}^{96}C_x p_r^x q_r^{96-x} = 0.0683,$$

whilst

$$\sum_{x=14}^{x=99} {}^{96}C_x p_a^x q_a^{99-x} = 0.1168 \quad \text{with} \quad \sum_{x=14}^{x=99} {}^{99}C_x p_r^x q_r^{99-x} = 0.9516.$$

Thus, if we take either value of n we shall have a scheme with properties close to those specified. In regard to these calculations it is important from the nature of the practical objectives of control rules not to fall into the trap of being too pedantic with respect to obtaining precise values of n and K. We could, for example, repeat the above calculations taking $K = 15$, we would then find that $n = 104$ for eqn (5.12) to hold and 106 for (5.13) in which case

$$\sum_{x=15}^{x=104} {}^{104}C_x p_a^x q_a^{104-x} = 0.0945 \quad \text{with} \quad \sum_{x=15}^{x=104} {}^{104}C_x p_r^x q_r^{104-x} = 0.9436.$$

We might even be tempted to take $K = 16$, in which case we would find that n is 112 and the above summations are then 0.1093 and 0.9533. In practical terms the differences in the errors of using the various schemes is trivial, but the last one would require 13 to 16 more samples than using $K = 13$. No doubt if we took $K = 17$ and $n = 120$ we might get values for the Type I and II errors even closer to 0.10 and 0.05. It is clearly important to be sensible about the choice of a control scheme, giving appropriate weight to the practical consequences of using the best scheme. The present example gives us the opportunity to ask the question what do we mean by best? Do we mean the test which minimizes n or do we mean the test which has closest values to the specified values of α and $1 - \beta$. The answer in practice is surely the former. We shall return to this question in a more realistic sense in a later chapter.

The arithmetic required to identify percentage defective schemes using the above method is, of course, greatly simplified by writing quite straightforward software. It is also simplified when $\alpha = 1 - \beta$ as the example below indicates.

Example 5.5

As in the previous example, let us take $p_a = 0.10$ and $p_r = 0.20$ but $\beta = 0.90$ rather than 0.95. Equation (5.11) then becomes

$$2x_\alpha \sqrt{B} = \ln\{(1 - p_a)p_r/[(1 - p_r)p_a]\} = 0.3134,$$

so that $K \geq 11$. With the above procedure we obtain $n = 80$ when $K = 12$. Either by using (3.7) or tables (for example, RND tables) we find

$$\sum_{x=12}^{x=80} {}^{80}C_x p_a^x q_a^{80-x} = 0.1004 \quad \text{and} \quad \sum_{x=12}^{x=80} {}^{80}C_x p_r^x q_r^{80-x} = 0.8994.$$

The method we have described is simple to use and requires only a little arithmetic. In addition to the limitations of the degrees of freedom in published tables of Z an advantage, is the ability to identify schemes for any values of α and $1 - \beta$, consider the following example.

Example 5.6

Cost and other considerations for a particular manufacturing process indicate that 4 per cent or less defective items is viable, provided that the frequency of detecting production batches with more than this percentage of defectives exceeds 1 in 12. If this proportion is 15 per cent or more, we require that on average only 1 in 50 batches of such material be received by customers.

To design a control scheme to satisfy these conditions $p_a = 0.04$, $p_r = 0.15$, $\alpha = 0.0833$ and $1 - \beta = 0.02$. Tables of the normal distribution give $x_\alpha = 1.383$ and $2.054 = x_{1-\beta}$. These parameters give $B \cong 0.1952$ and accordingly $K \geq 6$. We find the control rule with minimum n is $K = 6$ giving $n = 75$. The actual values of α and $1 - \beta$ are 0.0797 and 0.0234.

Let us use our examples to see whether there is an even easier way to determine rules with specific α and $1 - \beta$. Take eqn (3.28) which for sufficiently large n gives

$$\sum_{x=K}^{x=n} f(x; n, p) = 1 - F(K - 1; n, p) \cong 1 - \Phi[(K - np - 1/2)/\sqrt{(npq)}].$$

For the left-hand side of this equation to approximate to α when $p = p_a$ we require

$$K - np_a - 1/2 = x_\alpha\sqrt{(np_aq_a)}. \tag{5.14}$$

On the other hand, if this side of the equation is to be close to β then

$$1 - F(K - 1; n, p_r) \cong 1 - \Phi[(K - np_r - 1/2)/\sqrt{(np_rq_r)}]$$

and

$$\Phi[(K - np_r - 1/2)/\sqrt{(np_rq_r)}] = 1 - \beta,$$

giving

$$K - np_r - 1/2 = -x_{1-\beta}\sqrt{(np_rq_r)}. \tag{5.15}$$

We therefore need n and K which satisfy both (5.14) and (5.15), hence

$$n = [x_\alpha\sqrt{(p_aq_a)} + x_{1-\beta}\sqrt{(p_rq_r)}/(p_r - p_a)]^2 \tag{5.16}$$

and

$$K = (1/2) + np_a + x_\alpha\sqrt{(np_aq_a)}. \tag{5.17}$$

If we return to Example 5.3 and substitute $p_a = 0.05$, $p_r = 0.15$, $x_\alpha = 1.96$ and $x_{1-\beta} = 1.65$ into (5.16) it gives $n = 103$, whilst (5.17) gives $K = 9.98$. Thus to the nearest integer $K = 10$. Clearly, these values give a rule which very closely approximates to the one which emerged from the calculations of Example 5.3 by using an even simpler procedure. On the basis that one swallow does not make a summer perhaps we should take some other examples. Let us therefore consider Example 5.4. Substituting the values of $p_a = 0.10$, $p_r = 0.20$ etc., into (5.16) and (5.17) we have $n = 110$ and $K = 16$. For these values of p and equal Type I and II errors of 0.10, then $n = 82$ and $K = 12$. Finally, for Example 5.6 the normal approximation gives $n = 83$ and $K = 6$. At the level of sampling required for all of our examples, the additional inconvenience or cost of taking slightly larger samples than necessary seems a small price to pay having regard to the simplicity of this method.

A further and important reason for using the normal distribution is that evidently some of the expressions we have employed together with summations like (5.2) are not easy to manipulate from an algebraic point of view. It will become apparent that this is particularly the case in the development of decision rules which make the most effective use of test data. In so doing it becomes necessary to replace frequency functions like that of a binomial variate with close continuous approximations to them and which are more amenable to the methods of mathematics.

5.2 Proportion defective control charts

If the number of defectives in a sample is X we have seen that an efficient estimate of p, the proportion of defective items in the population, is $P_e = X/n$. There are some advantages to plotting values of this ratio on a chart against n rather than plotting values of X. There is a very arguable case for doing so, if for some reason the value of the sample size n might change between samples. An important feature of such a chart is that the position of its central line is fixed whatever the value of n. Let us consider $p_e = x/n$ and use Δp_e to denote the interval between consecutive values of p_e. For sample size $n\Delta p_e$ is $1/n$. If we write the frequency function of P_e as $f(p_e; n, p)$ then

$$f(p_e; n, p)\Delta p_e = {}^nC_x p^x q^{n-x},$$

so that for a particular value of P_e namely $p_e = x_1/n$

$$p(P_e \geq p_e; n, p) = \sum_{p_e=x_1/n}^{p_e=1} f(p_e; n, p)\Delta p_e = \sum_{x=x_1}^{x=n} {}^nC_x p^x q^{n-x}.$$

Since $E(P_e) = p$ and $\sigma^2(P_e) = pq/n$ it follows that

$$f(p_e; n, p)\Delta p_e \cong \sqrt{(n/2\pi pq)} \exp[-(p_e - p)^2 n/(2pq)] \mathrm{d}p_e. \tag{5.18}$$

Thus, to construct a control chart using P_e we would plot values of X/n on a chart with central line at p_a and the decision line drawn at $p_a + x_\alpha \sqrt{(pq/n)}$. Should the sample size for some reason vary between samples the sensible thing to do is to use standardized values of P_e on a chart. The central line would be at 0 whilst the decision line would be distance x_α above it. The value plotted on the chart for sample i with sample size n_i would be $(P_e - p_a)\sqrt{(n_i/p_a q_a)}$. This chart is called a standardized fraction defective chart. We shall shortly consider an example of the control of fraction defectives. For the moment, however, let us continue with consideration of the normal distribution in the context of the central limit theorem.

Much of industrial statistical quality control is concerned with controlling the mean m and standard deviation σ of a manufacturing process or those of batches of material produced by them. Suppose that the variate X with which we are concerned is continuous and that values of X are taken to assess the quality of a batch of material. For simplicity, assume for the moment, that σ is finite (as it usually is in practice) and remains constant from batch to batch. The batch mean m is to be monitored by the random sampling of n values of X. Let us also assume that the number of items in a batch is large. Thus for a particular batch we have n values X_i whose mean $\sum X_i/n$ is an estimate of the mean of the sampled batch. We accordingly assume that the values X_i are statistically independent of one another. This sample mean is the average of n statistically independent variates $(X_1, X_2, X_3, \ldots, X_n)$ each with frequency function $f(x; m, \sigma)$. If X is normally distributed we have seen that

$$f(x; m, \sigma) = \phi[(x - m)/\sigma]$$

and from Section 3.11 it follows that $\bar{X} = \sum_{x=1}^{x=n} X_i/n$ is $N(m; \sigma^2/n)$.

The central limit theorem gives the very important result that even if $f(x; m, \sigma)$ is not normal n does not have to be particularly large for the normal distribution with mean m and standard deviation $\sigma/\sqrt{n}$ to closely approximate the frequency function of this sample mean. What are the consequences of making this assumption in the design of control schemes? In doing so we find it a straightforward matter to produce control rules which require little mathematical expertise to formulate and extremely simple to operate. Thus from tables of the normal distribution function we find that $\Phi(3.09) = 0.9990$. In view of this, suppose that the acceptable mean value of X is m_a. The standard deviation of $\bar{X}$ is $\sigma/\sqrt{n}$. When testing is in control we shall only obtain values of this statistic greater than or equal to $m_a + 3.09\sigma/\sqrt{n}$ on average on 1 in 1000 occasions. This value of the sample mean is therefore so rare that if it occurs we would be inclined to the view that the mean of X has changed to a value $> m_a$. We find that $\Phi(3\sigma) = 0.9987$, hence from a practical point of view there is little difference between basing a decision on whether

$$\bar{X} > m_a + 3.09\sigma/\sqrt{n} \quad \text{or} \quad \bar{X} > m_a + 3\sigma/\sqrt{n}.$$

Likewise, tables give $\Phi(1.96\sigma) = 0.975$ whilst $\Phi(2\sigma) = 0.9773$. Accordingly, in very many applications of Shewhart charts it has become customary to use control lines placed at two or three standard deviations away from the acceptable mean level.

Whether we use two or three standard deviations depends, of course, upon the risks we are prepared to take of reaching wrong decisions.

In addition to simplifying the arithmetic of control charts in this way, the philosophy of not being too pedantic about the accurate determination of odds of making wrong decisions has considerable advantages. There are many applications of control charts, particularly in light engineering where sample values n can be of reasonable size and are sufficiently large to invoke the central limit theorem for statistics other than a sample mean. We shall see that to an acceptable level of approximation we can make the assumption that a number of control statistics $S(\mathbf{X})$ are normally distributed with mean $E[S(\mathbf{X})]$. Thus, if at acceptable quality $E[S(\mathbf{X})] = m_a(S)$ whilst the standard deviation of $S(\mathbf{X})$ $\sigma[S(\mathbf{X})] = \sigma_a(S)$, then the frequency function of $S(\mathbf{X})$, $f[S(\mathbf{x})]$ satisfies the expression

$$f[S(\mathbf{x})] \cong [\sigma_a(S)\sqrt{(2\pi)}]^{-1} \exp\{-[S(\mathbf{x}) - m_a(S)]^2/2\sigma_a^2(S)\}. \qquad (5.19)$$

If we want odds of the order of 1 in 1000 (say) of wrongly rejecting acceptable quality we do so only when

$$S(\mathbf{X}) \geq m_a(S) + 3\sigma_a(S). \qquad (5.20)$$

Let us take some examples to illustrate how simple this procedure is on the basis of appealing to the Central limit theorem, and how complicated things can become if we are over fastidious in the pursuit of absolute accuracy!

Example 5.7

Suppose we have a process where the proportion of rejectable items which can be tolerated is very small and we want the probability of rejecting batches to be rare if the proportion of defective items is less than 3 per cent. A large number of items is contained in a batch and testing is quick and cheap. Accordingly the manufacturer is prepared to sample 150 items from a batch to assess quality. If X is the number of defectives in such a sample what value of K should we take as a rejection criterion? What would be the consequences of using it for values of p greater than 0.03?

Solution

Making the assumption that the number of rejects in the sample will only rarely exceed $m(X) + 3\sigma(X)$ where $m(X)$ and $\sigma(X)$ are the mean and standard deviation of X

$$K = m(X) + 3\sigma(X) = np + 3\sqrt{(npq)} = 10.77. \qquad (5.21)$$

Thus, if the sample contains 11 or more defective items we reject the batch. We have seen that the actual probability of doing so is

$$\sum_{x=11}^{x=150} {}^n C_x p^x q^{n-x} = 1 - F(10; 150, p). \qquad (5.22)$$

The task of evaluating this summation for $p = 0.03$ and additional values of p is somewhat tiresome! However, values of $\Phi[(K - np - 1/2)/\sqrt{(npq)}]$ are easily obtained from tables of the normal distribution function. Values of the above sum and those using the normal approximation are given in Table 5.2.

Example 5.8

The mean value of a process is known not to vary once a production run has been set in motion. However, a machine fault can occur which can cause its standard deviation to change. It is known that measurement X used to monitor this is normally distributed. The standard deviation $\sigma(X)$ of X represents acceptable quality if it is ≤ 1.50. When this is so we want the risk of rejecting such quality to be small. We can take a sample as large as 25 values of X to control $\sigma(X)$.

Solution

To ensure rare rejection of acceptable batches we decide to design a '3σ' decision rule. For reasons given in Chapter 4, let us use $S_2(\mathbf{X})$ defined by eqn (4.18) to estimate $\sigma(X)$. For convenience let us write σ for $\sigma(X)$. The probability that $S_2(\mathbf{X})$ exceeds a given value K is obtained from (4.27). We have seen that $E[S_2(\mathbf{X})]$ is very close to σ and that the standard deviation of $S_2(\mathbf{X})$ is $[\sigma/\sqrt{(2n-3)}]$ and that the normal approximation to (4.27) is (4.26).

Using three standard deviation limits for a sample of 25 observations and $S_2(\mathbf{X})$ we have, for $\sigma = 1.50$,

$$K = \sigma[1 + 3/\sqrt{(2n-3)}] = 2.156. \qquad (5.23)$$

From tables of the distribution function (4.27) and those of the normal distribution function, we obtain the values shown in Table 5.3. Thus, in addition to the comparisons in Chapter 4, they confirm the utility of the central limit theorem in quality control situations where reasonably accurate approximations to probabilities are acceptable.

We have already remarked that it is not always easy to see how a particular problem can be converted or modelled into corresponding statistical questions. Let us therefore take a typical situation which can arise in an industrial context and examine in some

Table 5.2 Probability of rejecting a batch with 11 or more defectives in a sample of 150 for different values of p

p	$(K - np - 1/2)/\sqrt{(npq)}$	$\Phi(np - K + 1/2)/\sqrt{(npq)}$	$1 - F(10; 150, p)$
0.03	2.87	0.0020	0.0063
0.05	1.12	0.1305	0.1322
0.10	−1.22	0.8897	0.8940
0.15	−2.74	0.9970	0.9987

Table 5.3 Exact and normal approximations to $p[S_2(\mathbf{X}) \geq K]$

σ	K/σ	$\Phi[(\sigma - K)\sqrt{(2n - 3)}/\sigma]$	$(n - 3/2)K^2/\sigma^2$	$p[S_2(\mathbf{X}) \geq K]$
1.50	1.4376	0.0013	48.56	0.0021
1.75	1.2322	0.0584	35.68	0.0590
2.00	1.0782	0.2960	27.32	0.2898
2.25	0.9584	0.6122	21.59	0.6037
2.50	0.8626	0.8268	17.49	0.8270
2.75	0.7841	0.9306	14.45	0.9358
3.00	0.7188	0.9731	12.14	0.9782

detail its formulation into a statistical control procedure using the concepts we have described in Chapters 3 and 4. To do so we return to a situation which is very similar to Example 4.3.

Example 5.9

As in that example a production unit makes a component which is an essential part of a particular piece of electronic equipment. If it fails the equipment fails. An average component life m is acceptable from a marketing point of view if $m \geq m_a$. The product is being newly marketed and accordingly the values of m which are not acceptable are only roughly known. However, it is felt that a testing procedure should be introduced which ensures that batches with m less than $m_a/2$ will only rarely be used when assembling a piece of equipment. It is decided to sample n items and record the time to failure of each one.

Solution

If t is the time to failure of a particular part, then we have seen from eqn (4.19) that

$$f(t; m) = [1/m]\exp(-t/m).$$

From eqn (4.10) we have seen that the smallest value of the variance of an estimator of m is given by the expression,

$$\{-nE[(\partial^2/\partial m^2)\ln f(t; m)]\}^{-1}.$$

Since

$$(\partial/\partial m)\ln f(t; m) = [(t/m) - 1]/m \tag{5.24}$$

and

$$(\partial^2/\partial m^2)\ln f(t; m) = [1 - (2t/m)]/m^2, \tag{5.25}$$

it follows from this last equation that since $E(t) = m$ the minimum variance of an estimator of m is m^2/n. Furthermore, from eqn (4.11) it follows that if an estimator $S(t)$ of m exists with this variance then it satisfies the equation

$$\sum_{i=1}^{i=n} (\partial/\partial m) \ln f(t_i; m) = [S(t) - m]n/m^2.$$

Thus, using this result and (5.24) it follows that

$$S(t) = \bar{t} = \sum_{i=1}^{i=n} t_i/n. \tag{5.26}$$

Using the Neyman–Pearson result, the most powerful test for a given Type I error when $m = m_r$ is obtained when

$$\prod_{i=1}^{i=n} f(t_i; m_r) / \prod_{i=1}^{i=n} f(t_i; m_a) \geq k \qquad (k > 0).$$

We have seen that the test defined by this ratio is uniformly most powerful for values of $m > m_a$ if this ratio exceeds 0 when m is substituted in the above expression for m_r. We require

$$m_a^n \left[\exp \left(- \sum_{i=1}^{i=n} t_i/m_r \right) \right] \bigg/ m_r^n \left[\exp \left(- \sum_{i=1}^{i=n} t_i/m_a \right) \right] \geq k,$$

which gives

$$\sum_{i=1}^{i=n} t_i/n \geq [\ln(m_r/m_a) + (\ln k)/n](m_r - m_a)/m_a m_r = K,$$

and this is clearly a uniformly most powerful test. We therefore obtain the result that for a specified probability of wrongly rejecting batches with $m \geq m_a$, the statistic which has the highest probability of determining that m has become less than this value is $\bar{T}$. From Chapter 3 we have seen that the statistic

$$T = \sum_{i=1}^{i=n} T_i$$

is a generalized gamma variate with frequency function

$$f(t) = \{1/[\Gamma(n)m^n]\}t^{n-1} \exp(-t/m),$$

so that

$$p(\bar{T} \geq K) = p(T \geq nK) = [1/\Gamma(n)] \int_{nK/m}^{\infty} \omega^{n-1} \exp(-\omega) \, d\omega. \tag{5.27}$$

Table 5.4   Exact and approximate values of $p(T \geq K)$ for $m_a = 2500$ and $m_r = 1100$

m	$(K - m)\sqrt{n}/m$	nK/m	$\Phi[(m - K)\sqrt{n}/m]$	$p(\bar{T} \geq K)$
2500	−2.000	23.15	0.9773	0.9872
2200	−1.468	26.32	0.9289	0.9398
1900	−0.766	30.47	0.7782	0.7717
1600	0.200	36.18	0.4207	0.4000
1300	1.611	44.53	0.0534	0.0615
1100	2.980	52.63	0.0014	0.0041

Since the mean of $\bar{T}$ is m and $\sigma(\bar{T}) = m/\sqrt{n}$ as n becomes large, its distribution approaches that of a variate which is $N(m; m^2/n)$ and

$$p(\bar{T} \geq K) \cong 1 - \Phi[(K - m)\sqrt{n}/m]. \tag{5.28}$$

Suppose that $m_a = 2500$ units and that the sample size we can take is 35. We would therefore select 35 units at random from a batch, test each one to failure, and use their average life as the criterion for the acceptance or rejection of the batch. Suppose further that we would like 98 per cent of batches with average component life greater than 2500 units to be accepted. On the other hand, we want the control rule to very rarely pass those with mean life less than 1100 units.

Using the normal approximation to $\bar{T}$ we would then take

$$2500 - 2\sigma(\bar{T}) = K$$

so that $K = 1654$ and using eqns (5.27) and (5.28) we obtain the values given in Table 5.4.

Evidently the normal approximation gives values close to those of the exact probabilities.

5.3 Variate transformation

If the probability values given by a normal approximation with equivalent mean and standard deviation are not considered close enough to the exact probabilities, we can often improve them by transforming the variate under consideration. It is perhaps appropriate in the present context to point out that it is easy to fall into the trap of being over fastidious in the search for a normalizing variate transformation. In practice, the simpler the transformation the better. To illustrate this contention together with the possibilities of using transformed variates in the design of control schemes, let us consider one such transformation namely, X^λ using the result that

$$p(X \geq K) = p(X^\lambda \geq K^\lambda). \tag{5.29}$$

Can we find a value of λ for which the skewness of the distribution of X^λ as measured by eqn (3.9) is less than that of the distribution of X and for which its kurtosis is closer to that of the normal distribution, namely 3? If so, is the variate X^λ closer to normality than X?

Let us return to Example 5.9 to examine this proposition by considering $\bar{T}^\lambda$. From (5.26) it follows that

$$\mu_r(\bar{T}^\lambda) = (m/n)^{\lambda r} \Gamma(\lambda r + n) / \Gamma(n). \tag{5.30}$$

Stirling's approximation for $\Gamma(n + 1)$ given by eqn (4.17) gives the following close approximations to the mean and variance of the new variate

$$\mu_1(\bar{T}^\lambda) = m^\lambda[1 + \lambda(\lambda - 1)/2n] \tag{5.31}$$

and

$$m_2(\bar{T}^\lambda) = m^{2\lambda} \lambda^2 / n. \tag{5.32}$$

We can use (5.29) the recurrence relationship

$$\Gamma(\lambda r + n) = (\lambda r + n - 1)\Gamma(\lambda r + n - 1)$$

and tables of $\Gamma(1 + b)$ with $0 < b < 1$ (*Biometrika Tables For Statisticians*) to compute the skewness and kurtosis of $\bar{T}^\lambda$ for different values of λ. For example, if $n = 20$ the values of the two measures are given in Table 5.5. From it we see that the skewness and kurtosis of the distributions of variates with $\lambda = 0.50$ and 0.33 are very close indeed to those of a normal variate.

In view of these values we therefore ask does the use of the normal approximation

$$p(\bar{T}^\lambda \geq K^\lambda) \cong 1 - \Phi\{[K^\lambda - \mu_1(\bar{T}^\lambda)]/\sigma(\bar{T}^\lambda)\}$$

with $\mu_1(T^\lambda)$ and $\sigma(T^\lambda)$ given by the approximations (5.31) and (5.32) work? Taking $n = 35$ and $K = 1654$ as in Example 5.9 and $\lambda = 0.50$ and 0.33, Table 5.6 illustrates that when the need arises, we can get closer approximations to the probability that a variate exceeds a specified value by using the normal approximation to the distribution of a simple transformation of the original variate.

Table 5.5   Values of the skewness and kurtosis of $\bar{T}^\lambda$ for different λ

λ	Skewness	Kurtosis
1.00	0.2000	3.30
0.75	0.0785	3.10
0.50	0.0129	3.00
0.33	0.0000	2.98

Table 5.6   Approximate values of $p(T \geq K)$ given by eqn (5.33)

m	$z = K^\lambda - \mu_1(\bar{T}^\lambda)]/\sigma(\bar{T}^\lambda)]$		$1 - \Phi(z)$	
	$\lambda = 0.50$	$\lambda = 0.33$	$\lambda = 0.50$	$\lambda = 0.33$
2500	-2.1658	-2.2280	0.9848	0.9871
2200	-1.5305	-1.5540	0.9371	0.9399
1900	-0.7503	-0.7452	0.7735	0.7719
1600	0.2403	0.2541	0.4050	0.4000
1300	1.5564	1.5396	0.0598	0.0618
1100	2.7190	2.6397	0.0037	0.0040

An assumption of normality makes the design of control charts with specific Type I and Type II errors a very straightforward procedure. Suppose we consider the control of a population mean m of a variate which is $N(m; \sigma^2)$. A sample of n values X_i is taken at random from the population and we wish to design a scheme with Type I error of not more than α for $m \leq m_a$, and Type II error $\leq 1 - \beta$ for $m \geq m_r$. We have seen that the statistic we should use to monitor m is the sample mean

$$S(X) = \sum_{i=1}^{i=n} X_i/n$$

with the rule $S(X)$ greater than or equal to an appropriate value of K. Since the variates X_i are independent $N(m; \sigma^2)$ variates, $S(X)$ is $N(m; \sigma^2/n)$. We can determine the values of K and n required for the Type I and II errors specified from the two equations already described for the control of the proportion of defective items, namely

$$m_a + x_\alpha/n = K \tag{5.33}$$

and

$$m_r + x_{1-\beta}/n = K, \tag{5.34}$$

which give

$$n = [(x_\alpha - x_{1-\beta})\sigma/(m_r - m_a)]^2. \tag{5.35}$$

When n is known, K can be obtained from either (5.33) or (5.34). Thus if $m_a = 15$, $m_r = 25$, $\alpha = 0.025$ and $1 - \beta = 0.01$, then $x_\alpha = 1.96$, whilst $x_{1-\beta}$ is -2.33 so that $n = 26.5$ and $K = 19.56$. We have of course assumed that only the population mean m changes in the derivation of eqns (5.33) and (5.34). If the standard deviation of X_i changes as m changes so that when $m = m_a$, $\sigma = \sigma_a$ whilst it is σ_r when $m = m_r$ then,

$$n = [(x_\alpha\sigma_a - x_{1-\beta}\sigma_r)/(m_r - m_a)]^2.$$

Thus, suppose in Example 5.9 that in addition to a Type I error of 0.02 we can tolerate a Type II error of 0.01. What are the values of K and n which give a rule with these values? To the degree of accuracy implied by Table 5.3 an assumption of normality for the mean of nT values makes their calculation a simple matter. When the mean of this statistic is m its standard deviation is $m/\sqrt{n}$ and so

$$n = [(2.33m_r + 2.05m_a)/(m_a - m_r)]^2 = 30.16$$

and

$$K = m_a(1 - 2.05/\sqrt{n}) = 1566.$$

If we want to achieve the additional accuracy indicated by Table 5.5 the process for determining K and n is still quite straightforward. Suppose, for example, we take $\lambda = 0.33$, then the mean and standard deviation of $\bar{T}^\lambda$ are given by eqns (5.31) and (5.32) so that

$$m_a^\lambda[1 + \lambda(\lambda - 1)/(2n)] + x_\alpha \lambda m_a^\lambda/\sqrt{n} = K^\lambda \tag{5.36}$$

and

$$m_r^\lambda[1 + \lambda(\lambda - 1)/(2n)] + x_{1-\beta} m_r^\lambda/\sqrt{n} = K^\lambda, \tag{5.37}$$

so that n is given by obtaining the value of n obtained from equating these two expressions. Substituting $m_a = 2500$, $m_r = 1100$ and $\lambda = 0.33$ gives $n = 28.48$. Since n has to be an integer, we take it to be 28, we then have $K = 1628$. Based on the assumption of the normality of $\bar{T}^{0.33}$ these values estimate the Type I error of the test to be 0.02 and its Type II error to be 0.01. Using the actual d.f. of T given by eqn (5.27) we obtain the true value of these two errors to be 0.02 and 0.011.

This example again illustrates how use of the central limit theorem leads to a simple procedure to formulate tests with Type I and Type II errors close to those required in a particular situation. It also demonstrates how a simple transformation can be used to find tests with Type I and Type II errors which are very close indeed to those required when such close approximations are needed. In practice therefore, there are clearly situations where we can use normal approximations either

• to obtain tests which are easily formulated and whose accuracy is sufficient for practical purposes; or

• to indicate parameter values which may need subsequent refinement using the actual distribution function of a specific control statistic should additional accuracy be required.

A further illustration of the use of $\beta_1(X)$ and $\beta_2(X)$ and use of the central limit theorem with practically collected data is the following example.

Example 5.10

Bales of synthetic staple fibre were sampled to examine the quality of material dispatched to wool manufacturers. The data was collected by sampling a large number of bales, recording fibre lengths and subsequently noting those which carded satisfactorily and those which did not. The distributions of fibre lengths for bales found to be acceptable from this point of view and those which were not are given in Table 5.6. It was thought that a major reason for unsatisfactory carding of material in a bale was its mean fibre length. The data was accordingly used to determine this mean $m_a(X)$ from the data obtained for acceptable bales and $m_r(X)$ in the unacceptable ones. The information in the table was also used to design a control scheme to monitor this aspect of bale quality. The specification committee set up to supervise the scheme decided that it was prepared to recommend retesting 1 in 10 acceptable bales, if the control procedure used ensured that the average number of unacceptable bales dispatched to customers did not exceed 2.5 per cent. The control process should therefore have a Type I error of 0.10 and a Type II error of 0.025.

Solution

Clearly, from Table 5.7 the distributions of X, the fibre lengths, are not normally distributed. It was felt that rather than trying to fit a theoretical distribution to the data we should use the values of $f(x)$ in the table as defining the two frequency functions of X. We find that $m_a(X) = 1.417$ and $m_r(X) = 1.564$; $\sigma_a(X)$ the standard deviation of the distribution of acceptable quality is 0.2805 and $\sigma_r(X)$, that for unacceptable material is 0.4388. The values $\beta_1(X)$ and $\beta_2(X)$ for acceptable bale material are 1.35 and 5.26, whilst for that which is unacceptable, they are 1.00 and 4.82. For the Type I and II errors specified, an assumption of normality for the control statistic has $x_\alpha = 1.28$ and $x_{1-\beta} = -1.96$, giving $n = 84$ for the number of fibres to be tested per bale.

How close is the distribution of the mean fibre length likely to be to normality for such a sample size?

We can indicate an answer to this question by using eqn (3.48). Consider a variate X with m.g.f. $m(t)$ then if $m_S(t)$ is m.g.f. of the sum S of n independent such variates

$$m_S(t) = [m(t)]^n = [1 + m_2(t)t^2/2 + m_3(t)t^3/6 + \cdots]^n.$$

If we obtain the coefficients of t^2, t^3, and t^4 in the expansion of the right-hand side of this equation we obtain

$$m_2(S) = nm_2(X) \qquad m_3(X) = nm_3(X)$$

and

$$m_4(X) = nm_4(X) + 3n(n-1)m_2^2(X),$$

Table 5.7   Grouped data of fibre lengths in acceptable and unacceptable bales of staple fibre

Fibre length x	$f(x)$ Acceptable bales	$f(x)$ Unacceptable bales
< 1.10	0.016	0.036
1.10	0.108	0.101
1.20	0.200	0.109
1.30	0.124	0.125
1.40	0.108	0.109
1.50	0.092	0.101
1.60	0.040	0.093
1.70	0.044	0.060
1.80	0.024	0.040
1.90	0.024	0.044
2.00	0.008	0.065
2.10	0.008	0.028
2.20	0.004	0.008
2.30	0.000	0.012
2.40	0.000	0.008
2.50	0.004	0.016
2.60		0.012
2.70		0.008
2.80		0.000
2.90		0.008
3.00		0.004
3.10		0.008

so that

$$\beta_1(\bar{X}) = \beta_1(X)/\sqrt{n} \text{ and } \beta_2(\bar{X}) = 3 + [\beta_2(X) - 3]/n.$$

For acceptable quality we therefore have $\beta_1(\bar{X}) = 0.14$, $\beta_2(\bar{X}) = 3.03$ and for unacceptable quality their values are 0.10 and 3.02. As we know that the sample mean of any distribution with finite standard deviation approaches normality as the number of results it is based on increases, we conclude therefore that the probabilities generated by the normal approximation in this case should be enough to justify use of the rule in practice.

5.4 Randomness

An essential aspect of the operation of control charts is the assumption that the value X being plotted is indeed a random variate. When using simple Shewhart charts we need to check that its values fluctuate in a random fashion. We can do so by examining sequences of consecutive values of X. For example, we can analyse the lengths of runs R of a particular feature of X to see if it behaves in a fashion which accords with the assumption that X is a random variable. We can examine the way in which these features occur in sequences of consecutive values of X. What features can we use? For the control of the proportion of defective items it could be the distribution of the number of items sampled between the occurrence of one defective item and the next. Alternatively, it could be the distribution of consecutive values of X above and below its mean value $m(X)$. We can examine the number of 'runs up' and 'runs down' which occur in a sequence of values of X. Clearly, a succession of increasing values of X is a 'run up'. We can count the number of runs up and down to decide whether or not X is behaving randomly. It is obviously necessary to base such checks on a reasonably large number N of observed values of X. For the kinds of features just described, the distribution of R which we would expect when X is random is known to be approximately normal even when N is not particularly large. To illustrate this remark, suppose we decide to look at the number of runs up and down in a total of N values of X. Let n_1 be the number of values of X in 'runs up', so that $n_2 = N - n_1$ is the number of runs down. If n_1 and n_2 are both > 20 it can be shown that the distribution of R is approximately normal with mean

$$m = 1 + 2[(1/n_1) + (1/n_2)]^{-1} \qquad (5.38)$$

and variance

$$\sigma^2 = (m - 1)(m - 2)/(N - 1). \qquad (5.39)$$

To illustrate a test for randomness, let us take the data of Table 5.8 and give a plus sign to values of X_i which are greater than X_{i-1} and a negative one when X_i is less than X_{i-1}. A count of these signs gives $n_1 = 27$, $n_2 = 33$, $N = 60$ and $R = 36$. From eqns (5.38) and (5.39) we have $m = 30.7$ and $\sigma = 3.80$, so that using the normal approximation we find that

$$p(R \geq 36) = 0.0815,$$

which is clearly not significant.

There are, of course, other tests for randomness we can use such as the median test. Here we determine the value of X_i for which half of the values lie above it and half below. We can then determine the total number of runs above and below this median value and use (5.38) and (5.39) to compare it with the expected number of runs. Thus, for the data of Table 5.8 we find the median value of X_i is 7.83. The total number of runs for the data is $r = 24$. The normal approximation gives the probability of such a run or less as 0.0262 and that on average in 95 per cent of cases we should expect values of R to lie between 24 and 39.

A further test is to use the largest run length above or below the median value of X_i. Mosteller determined values of R, R_L which would only be exceeded in a run of N results with probabilities of 0.05, 0.01, and 0.001.

Table 5.8

i	X_i	Sign	i	X_i	Sign	i	X_i	Sign
1	3.09		21	8.67	+	41	6.47	+
2	12.89	+	22	6.95	−	42	5.13	−
3	8.79	−	23	3.03	−	43	5.73	+
4	6.80	−	24	8.96	+	44	4.44	−
5	3.85	−	25	4.37	−	45	3.20	−
6	8.58	+	26	5.82	+	46	10.10	+
7	9.73	+	27	5.86	+	47	7.42	−
8	11.59	+	28	7.83	+	48	6.93	−
9	12.45	+	29	9.27	+	49	10.08	+
10	7.01	−	30	6.28	−	50	9.39	−
11	4.16	−	31	7.23	+	51	8.29	−
12	9.40	+	32	6.32	−	52	12.28	+
13	8.09	−	33	9.07	+	53	11.93	−
14	9.70	+	34	11.54	+	54	4.13	−
15	9.65	−	35	12.74	+	55	12.85	+
16	6.58	−	36	7.28	−	56	9.19	−
17	8.25	+	37	4.74	−	57	8.32	−
18	10.50	+	38	9.84	+	58	8.41	+
19	8.26	−	39	7.32	−	59	6.95	−
20	3.28	−	40	3.94	−	60	4.34	−
						61	3.52	−

Some of the values he obtained are given in Table 5.9. The longest run of results above or below the median value of X_i in Table 5.8 is 7, which is well below the value given for $p(R \geq R_L) = 0.05$. On the whole, these three tests indicate that the values in Table 5.8 are just about randomly distributed.

When assessing the randomness of a set of results plotted on a Shewhart chart we are looking to check that the path they follow is in no way systematic, in particular that

- runs of results up and down are not longer or shorter than we would expect on the assumption of randomness;
- runs above or below the mean or median value of X are in accord with what we would expect in a random set of results;
- that the largest runs seen do not exceed specific values.

With regard to testing for randomness, or indeed any property of a variate X, we should note that the more tests we apply, the more likely the chance of reaching a wrong conclusion! This possibility is exacerbated by the fact that the tests we have just described are unlikely to be independent of one another. A sensible way to deal with this difficulty is to use more efficient control rules than the ones we have been

Table 5.9   Values of R_L with at least one value of R is $\geq R_L$ with probabilities 0.05 and 0.01

n	$p(R \geq R_L) = 0.05$ R_L	$p(R \geq R_L) = 0.01$ R_L
20	7	8
30	8	9
40	9	10
50	10	11
60	11	12
70	12	13

examining in this chapter. In particular, to use rules which are structured to detect non-randomness as well as testing for individual results which diverge from an expected value. We shall see how this can be done in the chapters which follow.

Before we leave our consideration of simple methods of control like Shewhart charts perhaps we should emphasize the advantages of simplicity in application and the calculations required to design them. These two considerations are particularly important when consideration has to be given to the capabilities of operatives or technicians required to operate them on a routine basis. We should also bear in mind that particularly in industrial applications there are many situations where testing is cheap and very accurate assessments of probabilities of right or wrong decisions are not needed. As an example let us consider the situation where we want to control the standard deviation σ of a process. Suppose that it is unrealistic to expect the individuals required to operate the control scheme in a production unit to use equations like (4.18) to routinely estimate σ. It is, however, possible for them to obtain the averages of a few observations. We can estimate σ by using the mean range obtained in a succession of small samples. Suppose in Example 5.8 that instead of using $S_2(X)$ we take a number of samples n_1 of size n_2 where n_2 is small. We can obtain the value of the range for each separate sample and use their mean to control σ. The number of values of X used in this procedure is $n = n_1 \cdot n_2$. In Example 5.8 we saw that $n = 25$ gave the test powers of Table 5.2. Clearly, using the mean range of a number of sub-samples of n observations is a less efficient use of data than using $S_2(X)$. Let us accordingly ask how much bigger than 25 does n have to be to achieve about the same power as the test of Example 5.8?

Solution

If we use X_{1i} and X_{si} to denote the largest and smallest values of X in sample i the statistic we propose to use is

$$S_3(X) = \sum_{i=1}^{i=n_1} (X_{1i} - X_{si})/n_1.$$

The probability distribution of sample ranges was obtained by E. S. Pearson (1941) as have the values of d_1 and d_2 for different n_2 such that

$$E(X_l - X_s) = d_l\sigma \quad \text{and} \quad \sigma(X_l - X_s) = d_2\sigma.$$

Some of these are shown in Table 5.10 together with values of K for which the probability $p(X_1 - X_s \geq K) = \alpha$. For convenience we have taken $\sigma = 1$. A comparison between the probabilities given in this table with those of a normal variate with equivalent mean and standard deviation, suggests that the use of this approximation for the distribution of the mean range of several samples of size n_2 will give probabilities which are not far from their actual values. Thus for $n_1 = 5$ a comparison between $p(X_1 - X_s \geq K)$ for the values of K in Table 5.10 are shown in Table 5.11.

The values given by the normal approximation for $n_2 = 4$ and $n_2 = 6$ are very similar to those above. Therefore,. if we decide to use $S_3(X)$ to monitor σ, we can estimate the value of $n = n_1 \cdot n_2$ needed for a test with the power we want, using the expression

$$p[S_3(X) \geq K] \cong 1 - \Phi[(K - d_1\sigma)\sqrt{n_1}/(d_2\sigma)]. \tag{5.40}$$

For specific Type I and II values of α and $1 - \beta$ at $\sigma = \sigma_a$ and $\sigma = \sigma_r$

$$n_1 = \{d_2[k_\alpha\sigma_a - k_{1-\beta}\sigma_r]/[(\sigma_r - \sigma_a)d_1]\}^2$$

and

$$d_1\sigma_a + (k_\alpha d_2\sigma_a/\sqrt{n_1}) = K.$$

Table 5.10   Values of d_1, d_2, and K for different n_2 and α when X is $N(0; 1)$

n_2	d_1	d_2	$p(X_1 - X_s \geq K)$			
			$\alpha = 0.01$	$\alpha = 0.05$	$\alpha = 0.95$	$\alpha = 0.99$
4	2.059	0.8798	0.43	0.76	3.63	4.40
5	2.326	0.8641	0.66	1.03	3.86	4.60
6	2.534	0.8480	0.92	1.25	4.03	4.76

Table 5.11   Values of $1 - \Phi[(K - d_1)/d_2]$ for values of K in Table 5.10 when $n_2 = 5$

α	$1 - \Phi[(K - d_1)/d_2]$
0.01	0.0270
0.05	0.0668
0.95	0.9620
0.99	0.9918

Table 5.12   Values of the normal approximation for $p[S_3(X) \geq K]$ when $n_1 = 7$, n_2 and $K = 4.89$

σ	$(K - d_1\sigma)\sqrt{n_1}/(d_2\sigma)$	$1 - \Phi[(K - d_1\sigma)\sqrt{n_1}/(d_2\sigma)]$
1.50	2.86	0.0021
1.75	1.43	0.0764
2.00	0.36	0.3594
2.25	−0.47	0.6802
2.50	−1.13	0.8708
2.75	−1.68	0.9535
3.00	−2.13	0.9834

Using these equations for $n_2 = 5$ we find the nearest integer value for n_1 to be 7 so that $n = 35$ and $K = 4.89$. Approximations to $p[S_3(X) \geq K]$ are shown in Table 5.12. If we were to use $n_1 = 4$, we find $n_2 = 9$ and $K = 4.35$ whilst for $n_1 = 6$ we have $n_2 = 6$ with $K = 5.28$. If we carry through the calculations equivalent to those obtained for Table 5.12 we find the values of $p[S_3(X) \geq K]$ are very close to those shown in this table. The values of n_2 selected for our example are the ones we would expect to use in practice. Evidently the larger the value of n_2 the less efficient it becomes as an estimate of σ. In practice, we should use values of n_2 close to 5.

This last example is informative from a practical point of view. Even in the present age of computerization and information technology there still remain many situations where those responsible for routinely operating and interpreting statistical control procedures find it easier to use simple statistics like $S_3(X)$ rather than $S_2(X)$. In addition, the need to increase sample sizes by about 10 as in this instance would not be a problem.

6
Better control rules

In Chapter 5 we have seen that Shewhart control charts are easy to design and use, particularly when the control statistic can be assumed to be normally distributed. There are two aspects of their design which we can now examine with some profit. The first is to recognize that basing their design upon the criteria of probabilities of right and wrong decisions may not be entirely appropriate in some industrial or clinical contexts. The second is a realization that these charts may not be making effective use of all of the information in sampled data with regard to moves away from acceptable control values. To illustrate these two remarks let us take a particular kind of industrial situation. Suppose we have a process, such as spinning a synthetic yarn which operates continuously over long periods of time. When it is operating satisfactorily the mean $\theta = \theta_a$ of a measured feature of the product which needs to be controlled remains constant. The technology of the manufacturing procedures is such that once the process is started the value θ_a will normally be maintained for long production runs. Circumstances can arise, particularly as the production run lengthens, when θ_a can change either gradually or quite suddenly. Accordingly, once the process is started samples are taken at regular intervals. The process is allowed to run until an out-of-control decision is made. Clearly, in situations of this kind material is not produced on a short-term batch basis. If R is the number of times the process is sampled up to the moment an out-of-control decision is reached, experience has shown it to behave in a random fashion. As already indicated in Chapter 1 the expected value $E(R; \theta)$ of this run length of a control scheme is its average run length (ARL) at quality level θ.

An important responsibility of clinical laboratories is the measurement of small concentrations of analyte, such as hormone or protein in samples of blood, urine, or saliva. Laboratories assay many patient samples every week, it is obviously very important to control levels of testing. This is frequently achieved by preparing a control pool of material which contains a fixed known concentration of analyte. Samples from the pool are inserted and tested in an assay at regular intervals.

Often the number of control samples which can be introduced in this way is restricted by cost, limitations of time, and frequently the need to prolong the life of the pool itself. For stable test procedures and diagnostic reasons it is clearly necessary to achieve long testing runs when procedures are in control and very short ones when they are not. In such circumstances it is evidently necessary to employ statistical procedures which make effective use of limited quantities of test material.

When the time interval between successive test samples taken from a continuously operating industrial process is a fixed value t, the total production time achieved between setting the control chart into operation and the out-of-control decision is Rt. Questions about the value of R is obviously equivalent to asking what production can we expect before the process is deemed to be out-of-control. Formulation of control schemes based on average run lengths means that using such a scheme, on average, the production run from the time of the first control sample tested is $tL(\theta_a)$ when testing is in control where $L(\theta) = E(R; \theta)$. If we assume a sudden change in θ from θ_a to RQL θ_r the production of out-of-control material will on average be between $tL(\theta_r)$ and $(t-1)L(\theta_r)$. There is, of course, an implied assumption in these last two remarks, namely that for any value of θ the probability that a given test will ultimately end with an out-of-control decision is 1. We shall see that we can easily establish that this is so. For many continuously operating chemical engineering processes, for example, an out-of-control decision means an expensive interruption in production. In clinical testing it implies the need to reassess test conclusions reached on patient samples. Both situations require analysis of the data to determine the time that the change in quality or testing level occurred. As already remarked since it is certain that an out-of-control decision will ultimately arise whatever the value of θ, we need the values of R to be high when processes are in control and as low as possible when they are not.

6.1 Average run length of a Shewhart chart

The frequency function $f(r)$ for the run length R of a Shewhart chart is easily established. If p is the probability that the control statistic $S(X)$ will lie outside the control line or lines, then the probability that $R = r$ is given by

$$f(r) = q^{r-1}p \quad r = 1, 2, 3, \ldots, \infty$$
$$= 0 \qquad \text{otherwise.} \tag{6.1}$$

Thus, whatever the value of $p(> 0)$, the total probability that $S(X)$ will indicate an out-of-control decision is

$$\sum_{r=1}^{r=\infty} f(r) = \sum_{r=1}^{r=\infty} q^{r-1}p = 1.$$

The expected value of r on p is

$$E(R; p) = \sum_{r=1}^{r=\infty} rf(r) = \sum_{r=1}^{r=\infty} rq^{r-1}p = \sum_{r=1}^{r=\infty} p[(d/dq)q^r]$$
$$= p\{(d/dq)[q/(1-q)]\} = 1/p. \tag{6.2}$$

6.2 Double- and single-sided control schemes

The line denoting the value K in a Shewhart chart of the kind described in the last chapter is called an action line for the obvious reason that action is taken when the value of the control statistic exceeds K. For the moment let us take the simple model where $S(X)$ is a sample mean, the process standard deviation σ of a single observation remains constant and X is a $N(m; \sigma^2)$ variate. The different situations which arise in practice with regard to the control of a process mean are as follows. There are circumstances where we need to control m to a specified target value m_t. The value of m_t can either take a single value or it can lie between specified upper and lower values of m, namely m_u and m_l. When we need to control m to a single value say $m_t = m_a$ the Shewhart control chart used has one action line. Such a chart is called a single-sided chart. If we wish to control m rigidly to m_a and need to detect both increasing and decreasing departures from m_a using random samples of size n, we would design a rule to detect changes with control lines at $m_a + k_1 \sigma / \sqrt{n}$ and $m_a - k_2 \sigma / \sqrt{n}$. The values of k_1 and k_2 being determined by the risks of right and wrong decisions we can tolerate, or the run lengths we need on average to achieve in the control of a process parameter such as its mean m. When m_t can take values between m_u and m_l the Shewhart chart would obviously have action lines at $m_u + k_1 \sigma / \sqrt{n}$ and $m_l - k_2 \sigma / \sqrt{n}$. Both of these charts are examples of double-sided schemes.

Example 6.1

Suppose the process mean for a normal variate X is in control if its mean m is equal to $m_a = 12$ and the sample mean of n observations is to be used as the control statistic. Suppose, further, that X has standard deviation $\sigma = 5.00$ which does not change when m changes. If m increases and becomes $\geq m_r'$ the process is judged to be out-of-control. If on the other hand m becomes $\leq m_r''$ with the difference $m_a - m_r''$ being equal to $m_r' - m_a$ then the process is also operating at an out-of-control level. To test for either eventuality a Shewhart control chart is to be used with control limits at $m_a \pm k\sigma_n$ where σ_n is the standard deviation of the control statistic X. The ARL of this scheme is easily obtained since

$$p(\bar{X} \geq m + k\sigma_n) = \int_{m_a + k\sigma_n}^{\infty} \phi[(x - m)/\sigma_n]\, dx = 1 - \Phi[(m_a + k\sigma_n - m)/\sigma_n]$$

and

$$p(\bar{X} \leq m - k\sigma_n) = \int_{-\infty}^{m_a - k\sigma_n} \phi[(x - m)/\sigma_n]\, dx = \Phi[(m_a - k\sigma_n - m)/\sigma_n]$$

so that from eqn (6.2) when $m = m_a$, $L(m_a) = E(R; m_a)$ is

$$L(m_a) = \{2[1 - \Phi(k)]\}^{-1}. \tag{6.3}$$

If m changes to m'_r, then from the two equations above

$$L(m'_r) = \{\Phi[(m'_r - m_a)/\sigma_n - k] + \Phi[(m_a - m'_r)/\sigma_n - k]\}^{-1}. \qquad (6.4)$$

Thus if we take $n = 25$ and $m'_r = 15$ and $k = 3.11$ then $\sigma_n = 1.00$, $L(m_a) = 500$ and $L(m'_r) = 2.19$.

Equations (6.3) and (6.4) can obviously be used to design a Shewhart chart with specified values of $L(m_a)$ and $L(m_r)$. For illustrative purposes suppose that sampling the process is sufficiently stable to permit just one sample a day, and sampling is expensive so that we want to keep n reasonably small. Can we design a scheme with $n < 25$? The scheme we have just considered gives an average of unnecessary adjustments of one every 16 months. In practice it is probably unrealistic to expect a process to run for this time without some kind of routine plant maintenance. It may therefore be more reasonable to design a scheme with a shorter in control run length. A scheme with an average interruption rate of 6 months could be a more practical proposition. Take $L(m_a)$ to be 180, and $L(m)$ equal to 3 when m increases to 15 or above, or if it becomes less than or equal to 9. What sort of reduction would we obtain in the size of the daily sample? From eqns (6.3) and (6.4) we want the values of n and k which satisfy the equations

$$\{2[1 - \Phi(k)]\} = 1/180 \quad \text{and}$$

$$\{\Phi[(m_r - m_a)(\sqrt{n}/\sigma) - k] + \Phi[(m_a - m_r)(\sqrt{n}/\sigma) - k]\} = 1/3.$$

From tables of the standardized normal d.f. we find $\Phi(2.77)$ is equal to 0.9972, so that $k = 2.77$. To find n we need only consider $\Phi[(m_r - m_a)(\sqrt{n}/\sigma) - k]$ with $m_r = 15$. Tables show that $\Phi(-0.4307) = 0.3333$ so that

$$[(3\sqrt{n})/5] - 2.77 = -0.4307,$$

which gives 15 for the nearest integer value of n. The scheme would therefore be a Shewhart chart with $n = 15$ rather than 25 and control lines at 15.58 and 8.42.

6.3 Choice between different schemes

We can design an infinite number of schemes with a specified ARL at acceptable quality level. It is informative at this point in our considerations of the formulation of control rules to ask the following question. How are we to choose between them? To indicate answers to this question let us consider Example 6.1 in a little more detail when the ARL is to be 180. Suppose that a testing level is acceptable as long as $n < 50$. If we use a sample of 20, 30, or 40 items, which of them should we take? We might intuitively feel that we should take the largest sample possible, say 40, $n = 40$ would surely be better than 20. This remark immediately poses the question what do we mean by better in the context of schemes designed on the basis of average run

Table 6.1   Values of $L(m)$ for two Shewhart chart schemes with $L(m_a) = 180$, $n = 20$ and $n = 40$

m	$L(m)$	
	$n = 20$	$n = 40$
12.0	180	180
12.5	107	75.2
13.0	43.7	21.5
13.5	18.7	7.8
14.0	9.1	3.6
14.5	5.0	2.1
15.0	3.0	1.4

lengths? For different values of m between 12 and 15 Table 6.1 gives the ARLs when $L(m_a) = 180$ for $n = 20$ and $n = 40$. The values of $L(m)$ in this table shed light on two practical considerations. The first relates to considerations of cost and effort. The cost of doubling the sample from 20 to 40 may well not be justified in order to achieve a difference of 1.6 for $m_r = 15$ or 9. Secondly, consider a situation where RQL may not be an altogether realistic concept and the need is to detect any change away from a specified target value as quickly as possible. From Table 6.1 the sample size $n = 40$ gives considerably lower values of $L(m)$ for all values of m greater than m_a. In these circumstances the higher sample size clearly gives a rule which has much better features than the one with $n = 20$. It is reasonable to suppose that we ought to be able to design a scheme with a similar and possibly better run length profile using control rules which utilize sampled data more efficiently than a Shewhart chart without the need to double the sample size.

With clinical testing in particular the need is often to reduce the expected value θ of a control statistic $S(X)$ as soon as θ moves away from the target value θ_a with minimal values of n. These two objectives can encompass a definition of better tests.

6.4 ARL profile

Where the detection of any changes in the value of a parameter θ is significant, it is evident that we can use $L(\theta)$ as a criterion on which to base the design of control rules. The ARL profile is the plot of $L(\theta)$ against θ. For schemes with a common value of $L(\theta_a)$ we can use their profiles to compare one with another. Better schemes will be those where $L(\theta)$ lies below the profiles of alternative schemes. If we can find a scheme or combination of schemes for which $L(\theta)$ lies below all other schemes it would be best in the sense implied here. We shall return to this definition of a best scheme in due course. Clearly, the eqns (6.1) to (6.3) imply that tests which are uniformly most powerful will give control schemes which are among the best which can be obtained. In this sense the control scheme we have just considered with $n = 40$ is uniformly better than that with $n = 20$.

6.5 More effective control charts

It is clear that simple Shewhart charts with action lines only do not make efficient use of available data with regard to processes which are likely to remain on target for long periods once they have been set up. In routine operation they frequently also rely on subjective judgements of whether or not plotted sample values are randomly distributed about the mean value of the control statistic $S(X)$. Let us consider a single-sided scheme. Suppose the plotted data are as shown in Fig. 6.1. The sample values of $S(X)$ indicate that the process is in control and that they are randomly distributed about the line representing the acceptable quality level.

If, however, the values of $S(X)$ resemble those of Fig. 6.2 the data strongly indicates that the parameter we want to control has in fact increased with time. No point, however, has risen above the action line, so that strict application of the control rule would not lead to an out-of-control decision. We might decide that the plotted points do not appear to be randomly distributed about the AQL value although the number of consecutive values of $S(X)$ above this value have not exceeded the value of the action line.

The decision as to whether the data relating to Fig. 6.2 indicates an out-of-control situation relies on whether or not the person charged with controlling the process concludes the data is no longer randomly distributed about the line representing AQL. This aspect of Shewhart charts is obviously unsatisfactory from his or her

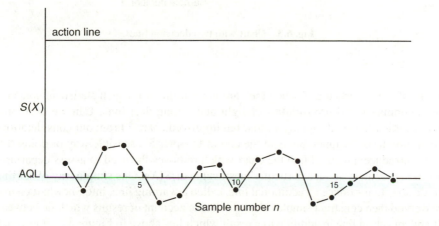

Fig. 6.1   In-control data.

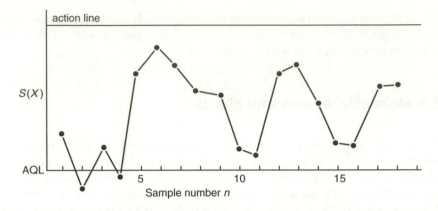

Fig. 6.2   Out-of-control data.

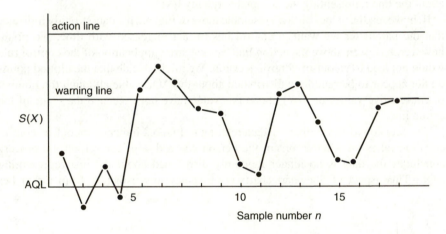

Fig. 6.3   Chart with two decision lines.

point of view, particularly when they are not familiar with such statistical concepts as randomness and probabilities of right and wrong decisions. Can we overcome these difficulties by designing similar but improved charts? From our considerations regarding tests for randomness at the end of Chapter 5 a sensible way of doing this is to introduce a method into the chart which removes the need to assess departures from randomness about a particular mean value of $S(X)$. One simple way of starting to do so is to introduce an additional line, called a warning line, into Shewhart charts. We could then construct simple rules which take account of results which lie between it and an action line in addition to a result which lies above it. Figure 6.3 shows such a line for the data of Fig. 6.2.

6.6 Charts with warning and action lines

We can investigate the advantages of such a scheme by considering a single sided control scheme with the following rule, take an out-of-control decision if two successive sample values $S(X)$ lie between a warning line and action line, or if a value of $S(X)$ lies above this last line. Let us call this rule Rule 1.

To obtain its properties use $p(r)$ for the probability that up to the rth sample no out-of-control decision has occurred and the rth value of $S(X)$ is below the warning line. Let p_0 be the probability that $S(X)$ takes a value below the warning line, p_1 that $S(X)$ lies between warning and action line, whilst $p_2 = 1 - p_1 - p_0$ is the probability that it lies above the action line then

$$p(r) = p_0[p(r-1) + p_1 p(r-2)]. \tag{6.5}$$

If $q(r)$ is the probability of a decision at the rth sample then for $r > 2$

$$q(r) = p_1(1 + p_2)p(r-2) + p_2 p(r-1) \tag{6.6}$$

whilst

$$p(1) = p_0 \quad \text{and} \quad q(1) = p_2 \tag{6.7}$$

and

$$p(2) = p_0(p_0 + p_1) \quad \text{with } q(2) = p_1(p_1 + p_2) + p_0 p_2. \tag{6.8}$$

From (6.7) and (6.5)

$$\sum_{r=1}^{r=\infty} q(r) = p_2 + p_1(p_1 + p_2) + p_0 p_2 + \sum_{r=3}^{r=\infty} q(r),$$

if we adopt the convention that $p(0) = 1$ then we can write

$$\sum_{r=1}^{r=\infty} q(r) = [p_1(p_1 + p_2) + p_2] \sum_{r=0}^{r=\infty} p(r)$$

$$= [1 - p_0(1 + p_1)] \sum_{r=0}^{r=\infty} p(r).$$

Using this convention we also have from (6.5) that

$$\sum_{r=1}^{r=\infty} p(r) = 1/[1 - p_0(1 + p_1)], \tag{6.9}$$

so that the probability that the rule ultimately terminates with an out-of-control decision is 1 since

$$\sum_{r=1}^{r=\infty} q(r) = 1. \tag{6.10}$$

The ARL of the rule under consideration is

$$E(R; p_1, p_2) = \sum_{r=1}^{r=\infty} rq(r),$$

so that from (6.6)

$$E(R; p_1, p_2) = p_1(p_1 + p_2) \sum_{r=0}^{r=\infty} (r+2)p(r) + p_2 \sum_{r=0}^{r=\infty} (r+1)p(r),$$

and using (6.5) we find

$$\sum_{r=0}^{r=\infty} rp(r) = p_0(2p_1 + 1)/[1 - p_0(1 + p_1)]^2$$

which using (6.9) gives

$$E(R; p_1, p_2) = L(p_1, p_2) = (1 + p_1)/[1 - p_0(1 + p_1)]. \tag{6.11}$$

In view of this result consider a single-sided scheme to control the mean of a variate which is $N(m; \sigma^2)$ when σ remains constant. We take a sample of n items and calculate their sample mean $S(X)$. Using Rule 1, if either a single value of $S(X)$ is greater than K or two successive values of $S(X)$ lie between W and K with $W < K$, we conclude the process is no longer in control. If m_a represents the AQL of m, $K = m_a + k\sigma/\sqrt{n}$ and $W = m_a + w\sigma/\sqrt{n}$ then, when the mean value of $S(X)$ is m,

$$p_1 = \Phi[k - (m - m_a)\sqrt{n}/\sigma] - \Phi[w - (m - m_a)\sqrt{n}/\sigma] \tag{6.12}$$

and

$$p_0 = \Phi[w - (m - m_a)\sqrt{n}/\sigma],$$

so that eqn (6.11) gives $L(m)$ as the ratio

$$\frac{1 + \Phi[k - (m - m_a)\sqrt{n}/\sigma] - \Phi[w - (m - m_a)\sqrt{n}/\sigma]}{1 - \Phi[w - (m - m_a)\sqrt{n}/\sigma] \cdot \{1 + \Phi[k - (m - m_a)\sqrt{n}/\sigma] - \Phi[w - (m - m_a)\sqrt{n}/\sigma]\}}. \tag{6.13}$$

Notice from eqn (6.13) that the value of $L(m)$ now depends upon the choice of three parameters, namely k, w, and n. In Table 6.2 where $n = 20$ we see that some

Table 6.2   Comparisons of the run length profiles of Rule 1 with a single-sided Shewhart chart for a $N(m; \sigma^2)$ variate when $n = 20$

m	Shewhart chart	Rule 1 $k = 2.54, w = 1.99$	Rule 1 $k = 3.10, w = 1.47$
12.00	180.0	180.2	180.0
12.25	97.2	94.8	82.2
12.50	54.9	52.2	42.6
13.00	20.0	18.2	14.1
13.50	8.7	7.7	6.3
14.00	4.4	3.9	3.6
14.50	2.6	2.4	2.4
15.00	1.8	1.7	1.9

values of k and w will evidently be better than others. It compares two rules of type 1 with the single-sided Shewhart chart with the same value of $L(m_a)$. The table shows that for all practical purposes the rule with $k = 3.10$ and $w = 1.47$ is uniformly better than the Shewhart chart for the range of values of m between 12 and 15.

To illustrate comparisons between the run length profiles of rules like Rule 1 for varying values of n and a higher $L(m_a)$, let us suppose the process under control can be sampled twice a day and that, as before, an average six-month production run is desirable when everything is in control. Take $n = 35$ and $L(m_a) = 360$ for variate X which is $N(m; \sigma^2)$ with $m_a = 12$ and σ constant at 5.

Let us consider the design of a scheme to control this target value against the background of the following observations on the scheme leading to equation (6.13). The rule consists of the imposition of one decision rule on to another, namely,

- take action if two points on a control chart occur above one line;
- take action if one point lies above another.

Suppose we need to design a scheme to detect changes above a target value m_a as quickly as possible. In addition if the mean of the process changes to m_r or more, then the ARL should be $\leq$ a specified value $L(m_r)$. Can we use the two rules just given to get a combined rule which achieves both objectives? It is reasonable to speculate that an appropriate choice of the first scheme ought to achieve the first objective, whilst the second should be obtained by positioning the action line in the right place on the chart. When $L(m_r)$ is small k will be large so that we can determine the value of w to obtain a given value of $L(m_a)$ by setting $k = \infty$ in eqn (6.13), so that

$$L(m) = \{2 - \Phi[w - (m - m_a)\sqrt{n}/\sigma]\}$$
$$\Big/ \{1 - \Phi[w - (m - m_a)\sqrt{n}/\sigma]\{2 - \Phi[w - (m - m_a)n/\sigma]\}\}$$

(6.14)

with

$$L(0) = [2 - \Phi(w)]/[1 - \Phi(w)]^2.$$

(6.15)

Table 6.3 Comparisons between the run length profiles of Rules 1 and their separate components

m	Rule 1		Two-point rule	Shewhart chart
	$k = 2.7764$	$k = 4.10$	$k = \infty$	
	$w = 2.319$	$w = 1.6063$	$w = 1.6063$	
12.00	360.0	360.1	360.1	360
12.25	147.5	120.4	121.3	151
12.50	66.4	47.6	48.0	68.6
13.00	16.6	11.3	11.82	17.9
13.50	5.80	4.80	4.92	6.28
14.00	2.70	2.80	2.96	2.92
14.50	1.68	2.10	2.29	2.34
15.00	1.25	1.76	2.08	1.23

From (6.15) we can obtain the value of w for which the first of these rules has a specified ARL when $m = m_a$, by solving the quadratic in $\Phi(w)$ of eqn (6.15). The value of k needed for a particular $L(m_r)$ can then be obtained from eqn (6.13). For illustrative purposes suppose that in addition to $L(m_a) = 360$ we want $L(m_r) = 1.75$. Equation (6.15) gives $w = 1.6063$, whilst eqn (6.13) gives $k = 4.074$. Thus, let us take Rule 1 with $w = 1.6063$ and $k = 4.10$ and compare its run length profile with those of the two component control rules with $L(m_a) = 360$, namely,

- two consecutive points above the line $m_a + w\sigma/\sqrt{n}$;
- one point above the line $m_a + k\sigma/\sqrt{n}$;

For the second rule $L(m_r)$ is given by

$$L(m) = \{1 - \Phi[k - (m - m_a)\sqrt{n}/\sigma]\}^{-1} \qquad (6.16)$$

which gives $k = 2.7729$. In our comparisons let us also include a second rule of type Rule 1 in order to indicate the advantages which might be gained by choosing w and k carefully. For this rule we take a value of k very close to the single sided Shewhart scheme. We find that $k = 2.7764$ and $w = 2.319$ gives $L(m_a)$ equal to 360. The run length profiles of these schemes are shown in Table 6.3.

The comparisons in this table have important implications which we shall consider in more detail in our search for control schemes which are close to the best that can be found. Values of $L(m)$ in the table indicate that there are advantages to be gained by investigating different combinations of k and w. It also indicates the advantages of combining decision rules in a way which uses the best features of both. The run length profile of the scheme with $k = 4.10$ and $w = 1.6063$ is uniformly better than the other Rule 1 shown in the table, except for values of $m > 14$. The differences between the $L(m)$ values for $m > 14$ will in many cases be trivial from a practical point of view. The rule with $k = 2.7764$ and $w = 2.319$ is, in fact, little better than a simple Shewhart chart. The significant feature of the table is the small differences (over the range of values of m considered) between the ARLs of Rule 1 with $k = 4.10$, $w = 1.6063$ and the two-point rule with $k = \infty$. We have seen that it is an easy matter to obtain w for a two-point rule. By considering pairs of values of k and w, we find that for a fixed value of $L(m_a)$ the schemes with the best profiles are those with values of $L(m)$ close to those of the two-point rule with the specified value of $L(m_a)$. The advantage of introducing an appropriately placed action line is to ensure low values of $L(m)$ in the region of $m = m_r$ without unduly affecting the influence of the profile of the two point rule in the region $m = m_a$. Thus to find optimal schemes for Rule 1 for specific values of $L(m_a)$ and $L(m_r)$ all we need to do is use eqn (6.15) to determine w and then use (6.13) to obtain k.

There are obviously other rules like Rule 1 which can be formulated. We could use a rule which requires a run of three results between a warning and action line rather than just two. Let us call this Rule 2. On the other hand, we can modify Rule 1 and formulate the rule that action is taken if two out of a sequence of N results lie between warning and action lines. For convenience call this Rule 3. Rule 1 is evidently a special case of Rule 3 when $N = 2$. Using the method we have described, we find that the run length $L(m)$ of Rule 2 is

$$L(m) = (1 + p_1 + p_1^2)/[1 - p_0(1 + p_1 + p_1^2)] \tag{6.17}$$

and for Rule 3 when $N = 3$ it is given by

$$L(m) = [1 + p_1(1 + p_0)]/[1 - p_0(1 + p_1 p_0)]. \tag{6.18}$$

Let us consider Rule 2 and compare its profiles with the one we have just been considering. From eqn (6.17) the numerator of $L(m)$ is

$$1 + \{\Phi[k - (m - m_a)\sqrt{n}/\sigma] - \Phi[w - (m - m_a)\sqrt{n}/\sigma]\}$$

$$+ \{\Phi[k - (m - m_a)\sqrt{n}/\sigma] - \Phi[w - (m - m_a)\sqrt{n}/\sigma]\}^2. \tag{6.19}$$

If we take this expression and the corresponding one for its denominator and substitute $k = \infty$, and follow the same procedure as before, we can find w from the equation

$$L(m_a) = \{3[1 - \Phi(w)] + \Phi^2(w)\}/\{1 - \Phi(w)\{3[1 - \Phi(w)] + \Phi^2(w)\}\}. \tag{6.20}$$

We can then obtain the value of k for a specified $L(m_r)$ using (6.18) or slightly less accurately by using eqn (6.19) to approximate $L(m_r)$, namely,

$$L(m_r) \cong 1 + \{\Phi[k - (m_r - m_a)\sqrt{n}/\sigma] - \Phi[w - (m_r - m_a)\sqrt{n}/\sigma]\}$$

$$+ \{\Phi[k - (m_r - m_a)\}\sqrt{n}/\sigma] - \Phi[w - (m_r - m_a)\sqrt{n}/\sigma]\}^2. \tag{6.21}$$

Using nothing more sophisticated than a pocket calculator and tables of the normal distribution function, we find $w = 1.0446$ and from (6.21) $k = 3.4861$ should give a

Table 6.4   Profiles of Rule 2 with $k = 3.4861$, $w = 1.0643$ and the constituent rules with $w = 1.0446$, $k = \infty$ and Shewhart chart with $k = 2.7727$

m	Rule 2	Constituent rule	Shewhart chart
12.00	360.2	360.2	360
12.25	106.5	109.3	151
12.50	41.2	41.5	68.6
13.00	10.5	10.9	17.9
13.50	4.78	5.1	6.28
14.00	3.07	3.7	2.92
14.50	2.26	3.2	2.34
15.00	1.69	3.0	1.23

value of $L(m_r)$ close to 1.75. Substitution of this value for k into the equation (6.18) changes w marginally to 1.0643. The values of $L(m)$ for the scheme $w = 1.0643$ and $k = 3.4861$ together with of the two constituent schemes with $L(m_a) = 360$ are shown in Table 6.4. When we take other values of w and k for rules of type 2 it is clear that the rule in Table 6.4 is for all practical purposes close to the best which can be achieved in the family of rules of type 2. The table also illustrates the important influence of the Shewhart chart to give values of $L(m_r)$ which in this case are less than 3. It also implies that Rule 2 is actually better than Rule 1 for changes in m close to m_a. Whether these differences are really significant from a practical point of view is debatable. However, one purpose of both rules is to include in the test an automatic assessment of the randomness of the control statistic being plotted. From this point of view the test which takes account of three results between warning and action lines is probably better than one which only takes two. Our comparisons also demonstrate that there is considerable choice with regard to the design of control rules. Consequently there is a need to devise appropriate criteria to identify better or best schemes.

The control rules we have just considered can obviously be extended to double-sided schemes in which warning and action lines are placed on either side of the target value or values, in order to detect both increases and decreases for example in m_a. Let us now consider a situation which arises in clinical testing where there are constraints on the amount of testing of control material which can be carried out. There are also limitations on retesting patient samples after an out-of-control decision has been reached. We can do so using the following example.

Example 6.2

The mean level of a pool of control material when a particular test procedure is under control is 10 units. The standard deviation σ of tests on the control material is constant and known to be 3 units. Samples from the control pool are introduced into routine testing once every day. If the mean of tests of samples taken from the pool increases by as much as 0.66 units, clinical considerations dictate that on average we need to know that such a change has occurred within 20 days. If, however, this mean increases by 3 units, patient samples will have to be reassessed. To minimize this retesting the average time to detect a change which is this large must only be 2 days. The technical complexities of testing and the costs of preparing a control pool are such that no more than 10 samples can be taken per day. Furthermore, the value of $L(m)$ when $m = m_a = 10$ should be as large as possible. Finally, senior laboratory staff require that the rule to be used should be one which is easy to understand and operate. Discussion on this last point leads to a decision that a rule of type 2 is sufficiently straightforward to introduce into the laboratory for daily operation by junior technical staff. The question which accordingly arises is, can we find values of k and w which satisfy the conditions required using the method just described?

Solution

We are told that the individual measurements X from a single sample can be assumed to be independent $N(m; \sigma^2)$. To simplify the arithmetic in the design of such a scheme and indeed its explanation to laboratory staff, we take $n = 9$ so that the control statistic, namely the sample mean is $N(m; 1)$.

To use the method implied from the conclusions drawn from Tables 6.2 and 6.3 we assume that the scheme with $k = \infty$ dominates the run lengths of our scheme for values of m in the region $m = m_a$. From eqn (6.17) it follows that

$$L(m) = G(w; m, m_a)/\{1 - G(w; m, m_a)\Phi[w - (m - m_a)]\} \qquad (6.22)$$

where

$$G(w; m, m_a) = 3\{1 - \Phi[w - (m - m_a)]\} + \Phi^2[w - (m - m_a)].$$

The solution to this equation when $m = 10.66$ and $L(m) = 20$ is $w = 0.8296$; solving the quadratic approximation (6.21) gives $k = 3.34$.

The value of $L(m_a)$ given by eqn (6.20) for $w = 0.83$ is 149. Our conclusions imply that this is the largest value of $L(m_a)$ which can be obtained when $n = 9$, $L(10.66) = 20$ and $L(m_r) = 13.0)$ is close to 2. The run length profiles shown in Table 6.5 indicate that this method of obtaining k and w leads with ease to the identification of schemes which for all practical purposes are about the best that can be achieved. It shows that if we need $L(m_r)$ very close to 2 then the maximum value of $L(m_a)$ is close to 137 if $L(m = 10.66)$ is to be 20. The table also illustrates that if we take $k = 3.70$ and $w = 0.83$, we obtain a scheme whose value of $L(m_r)$ is very close to 2 and has $L(m_a)$, nearly equal to what we believe to be the highest value we can get for the conditions imposed on the scheme. Whether we take $k = 3.34$ and $w = 0.83$ or instead take $k = 3.70$ depends upon which of these two run lengths is considered to be the most important one. The scheme in Table 6.5 with $k = 2.80$ and

Table 6.5   Run length profiles of control rules of type 2 with $L(m_r) \cong 2$ and $L(10.66) = 20$

m	$k = \infty, w = 0.83$	$k = 3.7, w = 0.83$	$k = 3.34, w = 0.83$	$k = 2.8, w = 0.7$
10.00	149	146	137	76.4
10.66	20.0	19.7	19.0	15.4
11.00	10.3	10.1	9.8	7.2
11.50	5.5	5.3	5.1	4.0
12.00	3.9	3.6	3.4	2.7
12.50	3.3	2.9	2.6	2.0
13.00	3.1	2.4	2.1	1.6

$w = 0.70$ is included to illustrate that the rules we have computed do indeed appear to be close to best.

Let us ask another question with regard to control charts with warning and action lines. How do their profiles change as n varies? Consider the following example.

Example 6.3

A manufacturing process operates continuously over a long period of time once it is set in motion. Its technology is such that it can confidently be expected to stay in control for about three months. The quality statistic X is normally distributed with mean $m_a = 5.72$ when the process is in control. If, however, m changes to a value above 7.00 we need to detect it quickly. There is, in fact, a critical value of the mean of X which if it is exceeded we need on average to find within one day's production. This value is 7.1313. The standard deviation of X does not change and is known to be 4.0. Management specify that although sampling is not particularly expensive they would like a scheme which minimizes daily testing.

After discussions with their statistician it is decided to use a type 2 rule with a run length of 120 days when the process is in control. If the critical value for the mean of X is reached then it should have a high probability of being detected within one day. It is suggested that samples of n values of X be taken at twelve hourly intervals so that the scheme should have $L(m_a) = 240$ and $L(m_r) = 2$. In view of the need to minimize the number of samples tested per day the statistician decides to examine schemes which sample the process every 6 hours and every 4 hours as well as the initially proposed 12 hours. It is also agreed that in view of the nature of the product we need to detect any change above $m_a = 5.72$ as quickly as we can.

Solution

When we determine w by taking $k = \infty$, we have seen that the addition of an action line reduces the value of $L(m_a)$ determined for this value of k. Let us determine w such that the scheme with $k = \infty$ has ARL $L(m_a) = L_1(m_a)$ sufficiently above that required allowing for the anticipated reduction in run length when the action line is added.

To illustrate this procedure, consider a scheme for sampling the process twice daily. We have seen from Table 6.5 that the difference between the values of $L(m_a)$ was 12, let us therefore in the present example examine what happens if we take $L_1(m_a) = 250$. The procedure for obtaining w, k, and n is then as follows. We first obtain w for $L_1(m_a)$ using the equation

$$3[1 - \Phi(w)] + \Phi^2(w) = \{\Phi(w) + [1/L_1(m_a)]\}^{-1}$$
$$= \{\Phi(w) + 0.0040\}^{-1}. \qquad (6.23)$$

The value $\Phi(w) = 0.8314$ satisfies this equation giving $w = 0.96$ and $L_1(m_a) = 250$. To compute k for $L(m_a) = 240$ we use

$$p_1 = \Phi(k) - \Phi(w)$$

in eqn (6.17) and have the quadratic in $\Phi(k)$

$$1 + [\Phi(k) - \Phi(w)] + [\Phi(k) - \Phi(w)]^2 = \{\Phi(w) + [1/L(m_a)]\}^{-1} \qquad (6.24)$$

from which we find $k = 3.59$. Finally, we obtain n by substituting $k - (m_a - m_r)\sqrt{n}/\sigma$ for k and $w - (m_a - m_r)\sqrt{n}/\sigma$ for w in eqn (6.24) and replacing $1/L(m_a)$ with $1/L(m_r)$. We then have that $n = 64$ gives $L(m_r) = 2.0$. These values give a scheme with run length in days of 120 when the process is operating satisfactorily, and one day when m moves to its critical value of 7.1313. Using this rule, we evidently require 128 values of X per day.

What happens if we decide to assess the plant on a 6-hourly basis rather than just twice a day? To obtain k and w we judge that we should take $L_1(m_a)$ to be 500. In doing so we find $k = 3.7529$ and $w = 1.1171$ gives $L(m_a) = 480$ and $L(m_r)$ equal to 4.08 with $n = 25$. This second rule is better than the first in the sense that the process is tested more frequently and the sampling level is quite a lot smaller. Obviously the increased sampling frequency will result in the chance of picking up certain changes in m earlier. Also the level of testing every 6 hours evens out the testing required from staff.

This last example is an important illustration of yet another element of choice in the design of control schemes, namely the sampling frequency. With it we can also show that there is an optimal frequency of testing. Consider what happens if testing is carried out on a 4-hourly basis. To obtain a scheme comparable to those above, we take $L_1(m_a) = 750$ and find $k = 3.8548$, $w = 1.2026$, $L(m_r) = 5.90$ and $n = 18$. Thus, the sampling level on each testing occasion is down to 18 but the daily testing load has increased to 108 from 100. In practice the choice of which of the last two schemes to use comes down to a decision as to whether it is more convenient to test

Table 6.6   Sample size and run length profiles of three rules of type 2 with common ARLs of 120 days at AQL

m	Sampling frequency		
	12 h $n = 64$ $k = 3.59$ $w = 0.96$	6 h $n = 25$ $k = 3.7529$ $w = 1.1171$	4 h $n = 18$ $k = 3.8548$ $w = 1.2026$
5.72	120	120	120
5.92	27.8	41.3	45.8
6.12	9.40	16.9	25.5
6.52	2.6	4.3	5.1
6.92	1.6	1.8	2.0
7.1313	1.0	1.0	1.0

25 values of the measure of control every 6 hours or 18 every 4 hours. The additional cost of 8 samples a day is probably neither here nor there.

   If we take account of the run length profiles of the three rules shown in the table, we see that when very small increases in m above m_a need to be detected quickly then the scheme we should use is the one with the largest sample size. If on the other hand we are reasonably ambivalent about changes in m in the range $5.72 \leq m \leq 6.12$ the test with a better profile is the second one.

   To illustrate further the advantages of rules using warning and action lines, let us return to Example 5.3 and the conclusions reached with it. The example related to an industrial process which is in control provided the proportion p of defective items is 0.05 or less, and has RQL of $p \geq 0.15$. The calculations carried through in Chapter 5 showed that to achieve a probability of rejecting acceptable quality as low as 1 in 40 and a probability of 1 in 20 of accepting rejectable quality, we would require a sample size of about 100 items.

   There are clearly many situations where such a large sample is unacceptable on a daily routine basis. Let us suppose that the process under control has the feature being assumed in this chapter that once set up and in control the proportion of defective items is unlikely to change for some time. Tests on the process made on one day can therefore be related to those taken in the next few days. When we can make this assumption we can use test data on a sequential basis using rules of the kind presently under discussion. If we do so and use run length as a criterion for the design of rules, can we reduce the level of sampling implied in Example 5.3?

   Let us therefore reconsider the example in some detail, in doing so we shall

- see that sampling can be substantially reduced;
- again illustrate that the method we have devised can be used to identify sampling schemes with reasonably optimal properties;
- show that we can utilize the central limit theorem to indicate the values of W, K, and n required for a particular control scheme.

Comments already made with regard to the difficulties in manipulating values of the d.f. of binomial variates clearly indicate the need for this last observation. Suppose we decide to sample the production process once a day and that we want $L(p_a) = 150$ when $p = p_a = 0.05$, and $L(p_r) = 2.5$ when $p = p_r = 0.15$.

Solution

In the notation of Chapter 3 write

$$f(x; n, p) = {}^nC_x p^x q^{n-x} \quad \text{and} \quad F(x_1; n, p) = \sum_{x=0}^{x=x_1} f(x; n, p)$$

then a type 1 rule using eqn (6.12) has

$$L(p_a) = [1 + p_1]/\{1 - p_0[1 + p_1]\}$$

where

$$p_0 = F(W; n, p) \text{ and } p_1 = F(K; n, p) - F(W; n, p), \tag{6.25}$$

so that

$$L(p) = [1 + F(K; n, p) - F(W; n, p)]$$
$$/\{1 - F(W; n, p)[1 + F(K; n, p) - F(W; n, p)]\}. \tag{6.26}$$

To obtain the value of W take the rule without an action line and inflate $L(p_a)$ to $L_1(p_a) = 160$. The value of p_0 for this scheme satisfies the quadratic in p_0

$$(2 - p_0)[p_0 + 1/L_1(p_a)] = 1 \tag{6.27}$$

which gives $p_0 = 0.91776$. To obtain a value for p_1 we use this value of p_0 and eqn (6.26) giving

$$p_1 = [p_0 + 1/L(p_a)]^{-1} - 1 \tag{6.28}$$

and $p_1 = 0.08175$. A search through tables of the binomial d.f. for values of n, w, and K to give these values of p_0 and p_1 is not particularly easy. Let us accordingly examine the possibility of using the central limit theorem to indicate the approximate position of appropriate warning and action lines together with the value of n. These can then be used to obtain more accurate values from tables of the binomial d.f. From eqn (3.28) we have

$$F(x_1; n, p) \cong \Phi\{[x_1 - np + 1/2]/\surd(npq)\}. \tag{6.29}$$

Thus, if we write $W = np + w\surd(npq)$ and $K = np + k\surd(npq)$

$$F(W; n, p) \cong \Phi\{w + 1/[2\surd(np_a q_a)]\}$$

and

$$F(K; n, p) \cong \Phi\{k + 1/[2\sqrt{(np_aq_a)}]\}$$

so that for $p_0 = 0.91776$ and $p_a = 0.05$,

$$w_1 = w + 1/[2\sqrt{(np_aq_a)}] = 1.39 \text{ and } k_1 = k + 1/[2\sqrt{(np_aq_a)}] = 3.27.$$

Use of approximation (6.29) to calculate $L(p_r)$ gives

$$p_0 \cong \Phi\{[w_1\sqrt{(p_aq_a)} + (p_a - p_r)\sqrt{n}]/\sqrt{(p_rq_r)}\} \tag{6.30}$$

and

$$p_1 \cong \Phi\{[k_1\sqrt{(p_aq_a)} + (p_a - p_r)\sqrt{n}]/\sqrt{(p_rq_r)}\} - p_0, \tag{6.31}$$

where p_0 is given by the approximation (6.30). In the present example we have $p_0 \cong \Phi(0.8484 - 0.2801\sqrt{n})$ and $p_1 \cong \Phi(1.9959 - 0.2801\sqrt{n}) - \Phi(0.8484 - 0.2801\sqrt{n})$. Using the expressions

$$K = np_a + k\sqrt{(np_aq_a)} \quad \text{and} \quad W = np_a + w\sqrt{(np_aq_a)}$$

we find our approximations indicate that n should be about 30, the value of K about 5 and $W \cong 3$. Taking these values as a guide, reference to tables of the binomial distribution function gives $n = 33$. Use of eqn (3.7) for $n = 33$, K equal to 6, and $W = 3$ gives $p_0 = 0.9192$ and $p_1 = 0.0798$, so that $L(p_a) = 144$ whilst $L(p_r) = 2.48$. Thus, to operate the control scheme, we would take samples of 33 items and conclude that the process is out-of-control only if we obtain more than 6 defectives in a sample, or if two successive samples each have 3 to 6 defectives in them.

Before leaving this example let us see what happens if we decide to operate a similar rule but take two samples a day rather than one. Thus, to have an ARL in days of 150 or thereabouts (since n, K, and W must be integers) when $p = p_a$ and 2.5 when $p = p_r$, we need $L(p_a) = 300$ and $L(p_r) = 5.0$. If we take $L_1(p_a) = 320$, and calculate p_0 when $k = \infty$ we find $p_0 = 0.9425$. The introduction of the action line using eqn (6.26) reduces $L(p_a)$ to 300 when $p_1 = 0.05725$. From our previous calculations we anticipate a reduction in n of more than $1/3^{rd}$ so that using tables of the binomial distribution we expect n to be in the region of 20. In fact we find that a rule with $n = 18$, $K = 5$, and $w = 2$ has $p_0 = 0.94187$ and p_1 equal to 0.05796 with $L(p_a) = 299$ and $L(p_r) = 5.08$. In this case there is no advantage to be gained by doubling the sampling frequency except that from time to time we shall pick up any changes which occur earlier.

In summary, there are two important aspects of this last example. The first is that it demonstrates the value of using rules which make more efficient use of data than a simple Shewhart chart, reducing the sample size required to much more reasonable levels than those implied by our examples in Chapter 5. It is clearly more convenient and cheaper to test samples of 20 or 30 items than 100. The second is the use of the central limit theorem to lead to the design of control schemes where the statistic being used has a distribution function which is awkward to handle. The example illustrates that the use of an appropriate normal approximation yields a method to determine parameter values which is quite straightforward. Furthermore, the values obtained using such an approximation will often be sufficiently accurate in their own right for many practical purposes. When this is so there is no need to use the refinements we have described, consequently the design of control rules is very simply achieved. Let us see whether this is also the case with regard to the control of process standard deviation. Consider the situation where individual measurements X are independent $N(m; \sigma^2)$ variate values. In view of our considerations in previous chapters suppose from a sample of n values of X_i all from this population we use

$$S_2(X) = \left[\sum_{i=1}^{i=n} (X_i - X)^2 / (n - 3/2) \right]^{1/2}$$

to estimate σ. Comparisons at the end of Chapter 5 show that the d.f. of $S_2(X)$ is close to normality when n is as small as 25. The prospect of using this approximation to design simply formulated control schemes for sample sizes of this order therefore appear quite promising. In addition, from eqn (4.27) and tables of the χ^2 distribution the design of schemes to control σ looks daunting! It is indeed difficult to decide where to start without the help of an approximating procedure. Let us take an example.

Example 6.4

Suppose we want to control σ_a and σ_r to the values 1.00 and 1.25 respectively, and that $L(\sigma_a)$ is to be 200 whilst $L(\sigma_r)$ is to be 4 or thereabouts. We decide to use $S_2(X)$ since we know that $E[S_2(X)]$ is very close to σ even for very small n.

Solution

To demonstrate the method let us still take a rule of type 2. To obtain a scheme which should be close to optimal and taking into account the calculations we have carried through in this chapter, let us inflate the ARL, required in the final scheme by 5 per cent. We therefore take $L_1(\sigma_a)$ to be 210 and, as before, obtain p_0 and p_1 which have $L(\sigma_a) = 200$. We find that $p_0 = 0.92857$ and $p_1 = 0.07116$. Using the approximation (4.26) we have for a sample size n likely to yield a scheme with $L(\sigma_a) = 200$.

$$p[S_2(X) \geq K] \cong 1 - \Phi[(K - \sigma)(2n - 3)/\sigma].$$

Substituting $K = \sigma_a[1 + k/\sqrt{(2n - 3)}]$ and $W = \sigma_a[1 + w/\sqrt{(2n - 3)}]$ we obtain that when $\sigma = \sigma_a$

$$p[S_2(X) \geq K] \cong 1 - \Phi(k) \quad \text{and} \quad p[S_2(X) \geq W] \cong 1 - \Phi(w),$$

so that $p_0 = \Phi(w)$ and $p_1 = \Phi(k) - \Phi(w)$ giving $k = 3.46$ and $w = 1.4652$. When $\sigma = \sigma_r$ we have from eqn (4.26) that

$$p_0 = \Phi[(w\sigma_a/\sigma_r) - (1 - \sigma_a/\sigma_r)\sqrt{(2n-3)}] \tag{6.32}$$

and

$$p_1 = 1 - \Phi[(1-\sigma_a/\sigma_r)\sqrt{(2n-3)} - k\sigma_a/\sigma_r] - \Phi[(\sigma_a/\sigma_r-1)\sqrt{(2n-3)} + w\sigma_a/\sigma_r]. \tag{6.33}$$

From these last two equations we find that $n = 25$ and for this value of n, p_0 is 0.4211 and $p_1 = 0.4976$, giving $L(\sigma_r) = 4.05$. We also have from the above equations that $K = 1.5047$ and $W = 1.2137$.

Returning to eqn (4.27) $v = 24$, $K^2(v-1/2)/\sigma_a^2 = 53.2068$ and the corresponding expression for W is $W^2(v - 1/2)/\sigma_a^2 = 34.6171$. Tables of χ^2 on 24 degrees of freedom give $p_1 = 0.07375$ and $p_0 = 0.9257$ when $\sigma = \sigma_a$, so that $L(\sigma_a)$ is equal to 178. When $\sigma = \sigma_r$, $K^2(v - 1/2)/\sigma_r^2 = 34.05$ and $W^2(v - 1/2)/\sigma_r^2 = 22.15$, giving $p_0 = 0.4297$ and $p_1 = 0.4813$ with $L(\sigma_r) = 4.11$. We can take the scheme with these run lengths to be good enough for practical purposes with regard to the features we require, or we can take the values of n, K, and W such as those just given together with tables of χ^2 to adjust to the actual values of K and W which have $L(\sigma_a) = 200$ and $L(\sigma_r)$ close to 4. If we do the latter and use an appropriate interpolation expression, we find in this case that for $n = 25$ we need to adjust K to 1.5362 and W to 1.2169, we then have $L(\sigma_a) = 201$ and $L(\sigma_r) = 4.36$.

These calculations give us an opportunity to emphasize a point of considerable practical significance in the design of control rules. With the development of new techniques it is important from time to time to stand back and view them in the context of what they are designed to do. The rules we have been examining in this chapter, and those we shall consider in Chapter 7 certainly have better features than those of simple Shewhart charts. That they are superior to the latter is not in doubt. With regard to the current example the values of K and W obtained on the normality assumption for $S_2(X)$ were easily found but gave a value of $L(\sigma_a) = 178$. The changes in K and W required to achieve a value of $L(\sigma_a) = 200$ are clearly minimal. We should surely ask, do we really need to modify the values of K and W obtained by the first method in order to adjust $L(\sigma_a)$ to precisely 200? It is clear from our considerations in this chapter and it will become even more apparent when we come to study cumulative sum charts in Chapter 7, that the values given for the ARL at AQL are critically dependent upon the assumptions made about the distribution of the control statistic being used. This aspect of the theory behind the design of rules needs to be fully recognized from the point of view of ease in design and use in practice. Whilst from a theorist's point of view it may be aesthetically rewarding to find ways of computing the values of constants like K and W which give $L(\sigma_a) = 200$ and have $L(\sigma_r) = 4.0$, we must seriously enquire on what realistic assumptions are these values based? Are they likely to be satisfied in practice? For instance, in the present example the assumption which is fundamental to the computed value of $L(\sigma_r)$ is that X is normally distributed. In practice this assumption will usually be an approximation. It is not difficult to show that even slight variations from normality for the distribution of X has quite a significant effect upon the value of $L(\sigma_a)$. We shall return to this point in the next chapter. We are not, of course, arguing that more efficient rules for controlling quality should not be used or developed. We can, however, be unjustifiably fastidious by undertaking interpolation calculations to achieve minor changes in the values of constants like K and W to obtain precise values of ARLs at AQL and RQL (say). Even small changes in the values of K and W which might be made for practical reasons can cause variations in the values of $L(\sigma_a)$ and $L(\sigma_r)$ equal to those which we may seek to remove through interpolation. Thus reference to tables of χ^2 shows that if instead of taking $K^2(v - 1/2)/\sigma_a^2 = 55.47$ and $W^2(v - 1/2)/\sigma_a^2 = 34.8016$ we took the values 56 and 35, then for $n = 25$ we have $L(\sigma_a) = 218$ and $L(\sigma_r) = 4.49$.

Before we leave this example let us examine whether we can use the normal approximations we have just described to lead us to a scheme designed to double the sampling frequency to twice a day with run lengths in days of 200 and 4. We know that this scheme will have $n < 25$ and the question is whether the normal approximations will work sufficiently well to obtain n, K, and W. Following through the procedure indicates that we can indeed do so. We find that n, $L(\sigma_a)$ and $L(\sigma_r)$ will be close to the values required if $p_0 = 0.95$ and $p_1 = 0.0499$. The value of n is 16 whilst $K^2(v - 1/2)/\sigma_a^2 = 42$ and $W^2(v - 1/2)/\sigma_a^2 = 25$ should give values of $L(\sigma_a)$ and $L(\sigma_r)$ close to those we need. In fact, we obtain $L(\sigma_a) = 404$ and $L(\sigma_r) = 8.58$. As a final example in this chapter let us consider the following situation.

Example 6.5

For a particular manufacturing procedure there is a need to control the standard deviation of its product to $\sigma = \sigma_a = 1.0$ units. The nature of the product is such that once set in motion and in control the plant is expected to run satisfactorily for about four months. The ARL when $\sigma = \sigma_a$ is therefore set by a specification committee to be 120 days. If, however, σ changes to a value of 1.35 units or more the committee decides that such changes should on average be picked up in four days. The statistician advising the committee suggests two alternatives. The first is to take all of the sample values obtained in a day and use an estimate of σ such as $S_2(X)$ to control it. The second is to sample more items but smaller samples in a day using the range of values in each sample to control σ. For a variety of reasons it was felt that this last suggestion would be preferable provided the number of individual items taken per day does not exceed 30. In the event the statistician decided to investigate both possibilities using a rule of type 1.

Solution

Following the notation of Chapter 5 we use X_1 and X_s to denote the largest and smallest values of X in a single sample of n items. In view of the observations made

Table 6.7   Run length profiles of schemes to control σ with Rules 1 and 2 with ARLs in days of 120 when $\sigma = \sigma_a = 1.0$ and $\sigma = \sigma_r = 1.35$

σ	$S_2(X)$		5 Ranges of 5	
	Rule 1 $n = 12, W = 1.294$ $K = 1.746$	Rule 2 $n = 13\ W = 1.166$ $K = 1.6652$	Rule 1 $W = 3.9638$ $K = 6.15$	Rule 2 $W = 3.3607$ $K = 6.15$
1.00	120	120	120	120
1.025	76.1	69.3	80.7	78.1
1.050	51.0	48.9	56.9	56.3
1.125	19.4	18.7	22.5	20.9
1.250	7.7	7.0	7.5	7.7
1.350	4.1	4.4	3.9	4.0

in this chapter let us take $n = 5$. Let us also take 5 such samples daily. For the specified daily run lengths the sample run lengths will therefore be 600 and 20. Using the method of inflating $L(\sigma_a)$ by 5 per cent and eqn (6.26), we find that $p_0(\sigma_a) = 0.95936$ and $p_1(\sigma_a) = 0.04055$ give the values of $L(\sigma_a)$ and $L(\sigma_r)$ required. Write $R = (X_1 - X_s)/\sigma$ and $f_n(r)$ for the frequency function of R when the sample size is n. Tables given by E.S. Pearson contain values of

$$F_n(r_1) = \int_0^{r=1} f_n(r)\,\mathrm{d}r.$$

From them we find $F_5(3.9638) = 0.9594$ and $F_5(6.15) = 0.9999$. These values of r_1 give $L(\sigma_a) = 596.5$. When $\sigma = \sigma_r = 1.35$, $r_1/\sigma_r = 2.9361$ when $r_1 = 3.9638$, and 4.5556 when $r_1 = 6.15$. Reference to the tables gives $F_5(2.9361) = 0.7695$ and $F_5(4.5556) = 0.9930$, so that $p_0(\sigma_r) = 0.7695$ and $p_1(\sigma_r) = 0.2285$ which give $L(\sigma_r)$ equal to 20.91. Thus the rule would be, take samples of 5 items 5 times a day, note the range of each sample, conclude the process is out-of-control only if two successive samples have ranges exceeding 3.9638 or one sample range is ≥ 6.15.

When we carry out the calculations for $S_2(X)$ we find that n is half the size of the scheme we have just described, for $n = 12$, $W = 1.294$ and $K = 1.746$, run length $L(\sigma_a) = 120$ and $L(\sigma_r) = 4.07$. The run length profiles of these two schemes together with those of the optimal schemes for Rule 2 with $n = 13$, $W = 1.1660$ and $K = 1.6652$ for $S_2(X)$ and $W = 3.3607$, $K = 6.15$ for 5 daily ranges of 5 items are shown in Table 6.7.

Part III

Control using cumulative sums

7
Really efficient use of test data

Let us now confine attention to the control of clinical testing procedures or industrial processes which once satisfactorily set in motion, control levels are normally maintained for long periods of time. As we have already remarked in much clinical testing it is necessary to utilize fully the information which is contained in quality control test data. We need to detect changes away from the target value of a parameter with the minimum amount of testing. Considerations of cost, diagnosis, the need to collect data for development and research, etc., also impose a need to minimize the frequency of unnecessary interruptions when testing is in control. We have seen that we cannot avoid the occasional wrong conclusion with regard to the values of population parameters when we use statistical control rules. In Chapter 6 we have demonstrated that with an appropriate choice of statistical technique and control statistic we can devise schemes with better run length profiles than those of straightforward Shewhart charts. The use of rules of types 1 and 2 produce schemes with better profiles than the latter with much smaller samples. With the object of achieving short runs of results when testing is out of control, and long runs when it is not, we now ask the following question. Can we devise testing techniques which are even better in these respects than those with warning and action lines? To indicate that we can do so, let us start with further consideration of rules like Rule 1 and Rule 2 of Chapter 6. For simplicity consider the rule, either take an out of control decision if one value of the control statistic lies outside the action line (or lines), or any u out of the last n samples have values between those represented by the warning and action lines.

To indicate the nature of this rule and its potential weakness in regard to achievable run lengths, let us assign scores to each sampled value of the control statistic according to whether it lies below the warning value, between this and the action value, or finally above it. Let us consider the possibilities of a scheme which assigns negative scores to control values of the first kind and positive ones to those of the second and third kind. Thus, if we devise an appropriate scoring system the occurrence of a value between warning and action line could be cancelled out if followed by a sufficiently long succession of values below the warning line. Consider in particular the following scheme. A sample of N items is taken at regular intervals, a score X_i is assigned to each sample which is

- $-a$ if a sample value is less than that assigned to the warning line;
- b if it lies between those of warning and action line;
- c if its value exceeds that assigned to the action line;
- $a, b,$ and c are all > 0 with $c > b$; and
- an out-of-control decision is taken at the rth sample if any of the sums

$$\sum_{i=0}^{i=j} X_{r-i} \geq h \quad (j = 0, 1, 2, \ldots, r).$$

The procedure is therefore to consider the scores X_i assigned to the last sample, the last two, three, four... samples etc. The run length of the scheme is the value of r at which one of these partial sums is $\geq h$. Page established that this procedure is equivalent to a sequence of sequential tests with decision interval h. We shall see that it is equivalent to plotting a control chart with horizontal boundaries at 0 and h, where each test of the sequence starts on the lower boundary. Testing is judged in control as long as members of each sequence which starts on this boundary ends on or below it without plotted values becoming equal to or exceeding h. The sequence ends when the plotted value $\geq h$. Using Page's reasoning we take samples of size n at regular intervals and assign score X_i to the ith sample and plot the cumulative sum

$$S_r = \sum_{i=1}^{i=r} X_i$$

on a chart and reach an out-of-control decision when from the chart we see that

$$S_r - \min_{0 \leq i < r} S_i \geq h. \tag{7.1}$$

We devise scores a, b, and c so that the mean path of the sums S_i is in a downward direction when testing is in control and changes to an upward direction when it is not. If we define $S_i(\max) = \max[S_{i-1}(\max) + X_i, 0]$ with $S_0(\max) = 0$ the sequential test we have just described is equivalent to plotting $S_i(\max)$ against i. The run length of the scheme is the value of i at which $S_r(\max) \geq h$. Figure 7.1, called a cumulative sum chart, illustrates an appropriate scoring scheme with plotted values of S_i. The individual test which comprises the sequence we have just described is called a Wald sequential test with horizontal decision boundaries.

We now ask whether we can find values of a, b, and c for which rules such as Rules 1 and 2 are equivalent to the operation of a sequence of Wald tests with boundaries 0 and h. We shall see that in particular instances this is indeed the case. In doing so we shall also see that the use of control rules based upon cumulative sums which fully utilize sampled data are likely to produce schemes with even better profiles than those illustrated in Chapter 6.

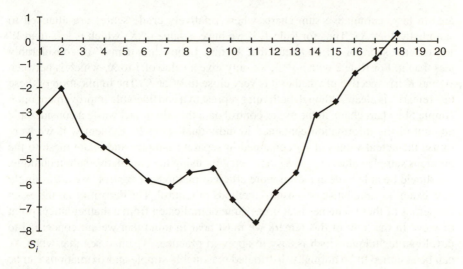

Fig. 7.1   Cumulative sum chart.

Take the rule which requires u of n sample values between warning and action lines or one point above this line. Clearly, $c \geq h$ and the following conditions also need to be satisfied,

- $(u-1)$ sample values between warning and action lines do not lead to an out-of-control decision;
- u values between warning and action lines and $(n-u)$ below the warning line must lead to an out-of-control decision, we therefore require that

$$(u-1)b < h \quad \text{and} \quad ub - (n-u)a \geq h; \tag{7.2}$$

- a sequence of $(u-1)$ sample values between warning and action lines together with $(n+1-u)$ below this line will result in this particular test having a total score ≤ 0, so that

$$(u-1)b - (n-u+1)a \leq 0. \tag{7.3}$$

The inequalities (7.2) give $b/a > (n-u)$ whilst (7.3) gives $(n-u+1)/(u-1) \geq b/a$. The schemes we are looking for require integer values of n and u to satisfy the inequality $(n-u+1)/(u-1) > (n-u)$ where $u \geq 2$. We therefore conclude that for this rule there are two schemes which are equivalent to Wald schemes with horizontal decision boundaries at 0 and h. These are any value of n when $u = 2$ or $u = n$. For the first of these two alternatives, $a = 1$, $b = n-1$ and $c = n$ with $h = n$. For the second, $a = n-1$, $b = 1$, and $c = n = h$.

The operation of a chart similar to Fig., 7.1 with an out-of-control decision taken when $S_r(\max) \geq h$ is a cumulative sum control chart. We see that the rules in Chapter 6 are, in fact, cumulative sum charts where relatively crude scores are attached to sampled values X_i. Thus for Rule 1, if we have a value of X_i which is less than W we assign a score of -1 no matter that X_i may be just less than W or considerably less than it. For Rule 2 with $n = 3$, we only give a value of 1 to X_i which is between W and K irrespective of whether it is very close to W or K. The implication of these last remarks is clear. Although both rules represent a considerable improvement upon simple Shewhart charts in the use of control data they do in fact waste a considerable amount of the information contained in individual samples. Evidently, if we were to use the actual values of X_i contained in separate samples and relate these to the previous sample values $X_{i-1}, X_{i-2} \ldots$ etc., by using the cumulative sum technique, we should be able to design even more efficient methods of control. We accordingly now examine cumulative or 'cusum' methods of control. The derivation of the exact properties of these schemes is at times quite complicated from a mathematical point of view. In the light of this remark we must bear in mind that we are concerned to develop a technique which is easy to apply in practice. We shall see that when X_i can be assumed to be normally distributed reasonably simple approximations can be devised for run length calculations which are sufficiently accurate for most practical applications.

With regard to the development of quite a few statistical techniques it is all too easy to fall into the trap of being too fastidious in the search for accurate probability values and the like, thereby losing sight of the fact that statistical analysis is in reality a practical tool. Some recent theoretical developments in the study of cumulative sum charts seem to have lost sight of this objective. In the development of a such a practical tool a prime concern must be to make the design of these control charts as simple as possible. We shall give simple expressions which can be used to do so. We shall also indicate a simple method for the construction of nomograms which with pencil and ruler can be used to design schemes.

7.1 The cusum technique

The essential features of the methods we shall now discuss are twofold. The first is the full utilization of information contained in a sample which relates to the value of the population parameter being controlled. The second objective is to relate this information to that contained in samples taken immediately before it.

Let us suppose we want to monitor the mean of a specific laboratory test. We do so by inserting control samples into the analytic procedure at specified intervals in time and monitoring the level of a constituent in each control sample. Samples are taken from a control pool for which the level of this constituent is known. The correct level of this constituent in the control pool m_a is known, X_i is the value obtained from the analysis of the ith sample taken from it. Whilst testing is in control the mean value of X_i is m_a. Testing is judged out of control when a change in this mean value occurs. For simplicity let us begin with situations where the value of σ, the standard deviation of X_i is known. It is also known that σ does not change in value when testing ceases to be in control; that is, when the mean of X_i changes to m_r. If we compute the sum

$$S_n = \sum_{i=1}^{i=n} (X_i - m_a) \tag{7.4}$$

and progressively plot S_n against n, the resulting chart is the cumulative sum of the differences between each control value X_i and its 'target' value m_a. As long as testing is on target the value of $E(X_i) = m_a$ and the path of the chart will fluctuate about a horizontal line. If it changes, the path direction will change. The frequency with which the difference $(X_i - m_a)$ exceeds 0 increases as the mean of X_i increases from m_a to m_r. The path of the chart accordingly begins to slope upwards. It will, of course, slope downwards if the mean test level of X_i drops below m_a. When the mean level of X_i is m_r from the beginning of sampling, then

$$E(S_n) = n(m_r - m_a). \tag{7.5}$$

If the mean of X_i suddenly changes from m_a to m_r the path of the cusum changes from a horizontal direction at the point of change. When the change occurs between the jth and $(j + 1)$th sample, then from this point onwards values of S_n fluctuate about the line $(n - n_j)(m_a - m_r)$. Changes in chart direction will, of course, arise from the random nature of X_i. We obviously require a test to distinguish directional changes attributable to these separate causes. Figures 7.2 and 7.3 indicate a set of values of $(X_i - m_a)$ plotted on a Shewhart chart and the same data plotted cumulatively. Figure 7.3 indicates the possible point of change much more clearly than Fig. 7.2.

7.2 Visual aid

Used with care, this feature of cusum charts can be effectively exploited. Visual inspection can indicate both changes in test level and the approximate point at which it occurred. This second aspect of plotted cumulative sums can evidently be important with regard to identifying causes of change. The cumulative sum technique can therefore fulfil the dual role of control chart and investigational tool.

The visual appearance of a chart of S_n plotted against n is very sensitive to changes in the ratio of vertical and horizontal scales used on the chart. If a chart is being used as a device to investigate possible reasons for changes in test or process level as well as for purposes of control, it is important to relate vertical and horizontal scales so that the directional changes due to random fluctuations in X_i are minimized, and emphasis is given to changes in slope associated with significant changes in the level of the population parameter under control. If we use ω for the ratio of the unit scale of n to that of the unit scale of S_n, we find that reasonable visual discrimination is achieved when $\omega = 2\sigma$ where σ is the standard deviation of X_i. Thus if the mean of X_i changes from m_a to $m_r = m_a + 2\sigma$ then after the point of change the path of S_n

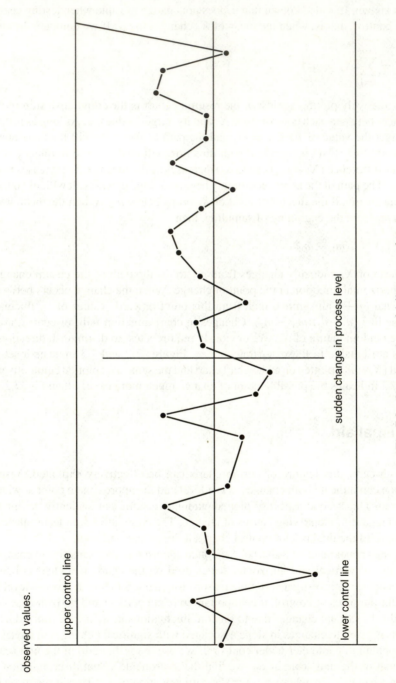

Fig. 7.2   Shewhart chart of daily test results.

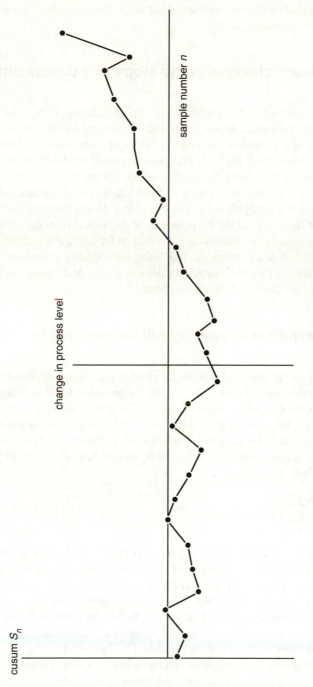

Fig. 7.3 Cumulative values $(x_i - k)$ plotted from test results of Fig. 7.2.

defined by eqn (7.4) will cease to fluctuate about a horizontal line and will begin to fluctuate about a line at 45° to it.

7.3 Significant changes in the slope of a cusum chart

To devise a control rule with specified average run lengths $L(m_a)$ and $L(m_r)$ we can use a simple geometric device suggested by Barnard (1959). This involves the progressive use of a V-mask on the chart as values of S_n become available. It is used in the manner indicated by Fig. 7.4. The mask is placed on the chart so that the line OP is horizontal and the point P coincides with the last plotted value of S_n. If the path of the cusum chart crosses either limb of the V-mask, conclude that the parameter being controlled has changed and is out of control. In the diagram line OP bisects the angle 2α of the mask, whilst the point P is at distance d from the vertex of the V. This distance, called the lead distance, is located on OP using the vertical scale of the cusum chart. Thus $d = 1$ corresponds to P being unit distance in terms of the vertical scale from O. We shall see that specified values of $L(m_a)$ and $L(m_r)$ can be obtained with an appropriate choice of the lead distance and α.

7.4 An alternative to plotting full cusum charts

Cusum control can be operated without the need to plot the whole chart when this is regarded as unnecessary. To take an example, suppose we wish to design a scheme to detect an increase in $E(X_i)$ from m_a to m_r. The procedure is to cumulate sample values only when values of X_i greater than $m_a + (m_a - m_r)/2$ are obtained. Suppose the first occasion that this happens is when $i = j$. From this point on, values of $X_i - (m_a + m_r)/2$ are cumulated either until the sum returns to zero or less or is $\geq$ to a value h. Thus, if we write

$$S_1(j) = \sum_{i=t}^{i=j+t} [X_i - (m_a + m_r)/2],$$

we calculate $S_1(j)$ until either $S_1(j) \leq 0$ or $S_1(j) \geq h$. If the first of these happens cumulation ceases until the next value of $X_i \geq (m_a + m_r)/2$ occurs, at which point the next series of values of $S_1(j)$ commences. Calculation of $S_1(j)$ continues until it becomes ≤ 0 or $\geq h$. This procedure is continued, cumulation being 'triggered' until the first occasion that cumulation leads to a sum $\geq h$. When this happens, conclude testing or production is out of control. The value of h is called the decision interval. A diagrammatic representation of the procedure is illustrated in Fig. 7.5.

To detect negative as well as positive changes in the expected value of X_i a second cumulative sum is computed simultaneously with $S_1(j)$. Suppose we want to detect changes in the mean of X_i outside the range $m_a \pm \Delta$ where Δ is equal to $m_r - m_a$

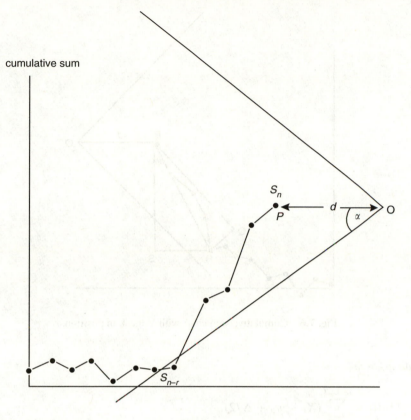

cumulative sum

Fig. 7.4   Cumulative sums $(x_i - m_a)$ plotted against sample number n.

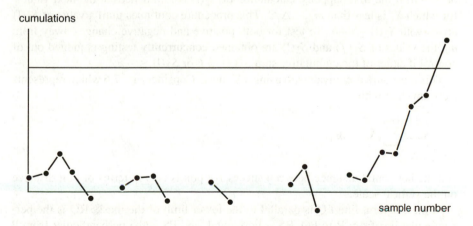

cumulations

sample number

Fig. 7.5   Cumulated sums of $(x_i - m_a - \Delta/2)$ plotted only when required.

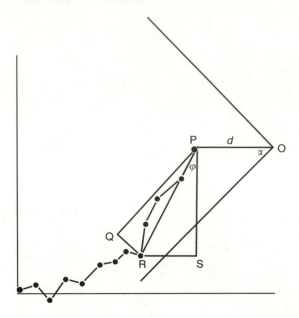

Fig. 7.6   Cumulative sum chart with V-mask in position.

with $m_r > m_a$

$$S_2(\mathrm{l}) = \sum_{i=\mathrm{T}}^{i=\mathrm{l}+\mathrm{T}} (X_i - m_a + \Delta/2),$$

l being the first occasion that $X_i < m_a - \Delta/2$ compute $S_2(\mathrm{l})$ until it becomes ≥ 0 or $\leq -h$. If the first happens, calculation of $S_2(\mathrm{l})$ recommences at the next value i for which X_i is less than $m_a - \Delta/2$. This procedure continues until a value of $S_2(\mathrm{l})$ arises with $S_2(\mathrm{l}) \leq -h$. To test for both positive and negative changes away from m_a the values of $S_1(j)$ and $S_2(\mathrm{l})$ are obtained concurrently testing is judged out of control if either of the cumulative sum $S_1(j) \geq h$ or $S_2(\mathrm{l}) \leq -h$.

This procedure is equivalent to using a V-mask. Consider Fig. 7.6 which represents a cusum chart with

$$S_n = \sum_{i=1}^{i=n} (X_i - m_a)$$

and the horizontal distance between successive points is ω in terms of unit distance on the vertical scale.

In the diagram, line PQ is parallel to the lower limb of the mask, RQ is the perpendicular line from R to PQ, RS is horizontal, and PS is the perpendicular from P to RS. Clearly, the $\angle SPQ = (\pi/2) - \alpha$, if we write $\angle RSP = \phi$ and P is the point S_n

whilst R is S_{n-j} then

$$QR = PR \cos(\phi + \alpha) = (S_n - S_{n-j}) \cos \alpha - j\omega \sin \alpha. \tag{7.6}$$

The path of plotted points on the chart crosses the lower limb of the mask when $QR \geq d \sin \alpha$. When this is so, it follows from eqn (7.6) that

$$\sum_{i=n-j+1}^{i=n} (X_i - m_a - \omega \tan \alpha) \geq d \tan \alpha. \tag{7.7}$$

A similar argument shows that the upper line of the V is crossed when

$$\sum_{i=n-j+1}^{i=n} (X_i - m_a + \omega \tan \alpha) \leq -d \tan \alpha. \tag{7.8}$$

The use of a V-mask on a chart which cumulates values of $(X_i - m_a)$ is therefore the same as forming both sums (7.7) and (7.8), and concluding that the expected value of X_i is off target when the first sum is $\geq d \tan \alpha$, or the second is $\leq -d \tan \alpha$. This is equivalent to the simultaneous operation of two single-sided decision schemes where the same values of the variate X_i are used for both. The decision boundaries of each scheme consist of two parallel horizontal lines distance $d \tan \alpha$ apart. For the first, the lower boundary is taken as zero and the upper one is used as the decision boundary; $\omega \tan \alpha$ is subtracted from each value of $(X_i - m_a)$ and as soon as a positive difference occurs this and subsequent deviations of $(X_i - m_a)$ from $\omega \tan \alpha$ are added together. These cumulated sums are plotted until their path crosses either boundary and, as already described, a decision that testing or a specific process is off target is made on the first occasion that the upper boundary is crossed. If the lower boundary is crossed, plotting is terminated until the next value of $(X_i - m_a)$ exceeds $\omega \tan \alpha$. The second scheme is similar to this, but here the upper boundary has the value 0 whilst the lower one is the decision boundary. A value $\omega \tan \alpha$ is now added to each value of $(X_i - m_a)$ and cumulation commences when a negative value of $(X_i - m_a + \omega \tan \alpha)$ is obtained.

7.5 The ARL of a V-mask

It has become customary to call the value $\omega \tan \alpha = k$ the reference value of a particular scheme whilst $d \tan \alpha = h$ is its decision interval. For convenience let us write $(X_i - m_a) = Y_i$ so that (7.7) and (7.8) become

$$S_1(j) = \sum_{i=t}^{i=j+t} (Y_i - k) \quad \text{and} \quad S_2(l) = \sum_{i=T}^{i=l+T} (Y_i + k).$$

Consider the two sums $S_1(m)$ and $S_2(n)$ where, $0 < S_1(j) < h$ with $(0 \leq j \leq m)$ and $0 > S_2(l) > -h$ with $(0 \leq l \leq n)$.

If the cumulative sums of the two schemes lie between their respective boundaries immediately after the rth sample is drawn, $S_1(m)$ and $S_2(n)$ will represent them if $m + t = n + T = r$. Consider what happens if $S_1(m + 1)$ becomes $\geq h$ so that an indication that $E(X_i)$ is off target is given by the first scheme. We have

$$S_1(m + 1) = S_1(m) + Y_{m+1} - k \geq h$$

so that $Y_{m+1} \geq h + k - S_1(m)$. The cumulative sum of the second scheme is then,

$$S_2(n + 1) = S_2(n) + Y_{m+1} + k \geq S_2(n) - S_1(m) + h + 2k$$

and

$$S_2(n) - S_1(m) + h + 2k = \begin{cases} S_2(t - T - 1) + 2(m + 2)k & t > T \\ h + 2(n + 2)k & t = T \\ h + 2(n + 2)k - S_1(T - t - 1) & t < T. \end{cases}$$

so that $S_2(n + 1) > 0$. Similar reasoning shows that $S_1(m + 1)$ must be less than zero when $S_2(n + 1) \leq -h$. When the cumulative sum immediately after the rth sample of a scheme lies between its boundaries and the cumulation of the other has ceased, a similar argument shows that only one decision boundary of the two schemes can be crossed at the next sample and that the cumulative sum of the other cannot lie between its boundaries. It follows that a cumulation which ends on or above the upper boundary of the first scheme cannot cut short a cumulation which would have ended on or below the lower boundary of the second and vice versa.

Consider the situation when only the cumulative sums of the first scheme are plotted. If $L_1(m)$ is the ARL of this scheme when $E(X_i) = m$ the expected number of times its upper boundary is crossed when N samples are taken is clearly $N/L_1(m)$. Cumulations for these will not be affected if we now plot the same results for the second scheme. Those obtained for this scheme which crosses the boundary $-h$ only occur at points where cumulations for the first scheme have ceased because its lower boundary has been reached or crossed. If $L_2(m)$ is the ARL of the second scheme the expected number of times its decision boundary is crossed is $N/L_2(m)$. In view of the complete dependence between the two schemes it therefore follows that the ARL $L(m)$ for the simultaneous operation of both must satisfy the relationship

$$N/L(m) = N/L_1(m) + N/L_2(m)$$

and so it follows that

$$L(m)^{-1} = L_1(m)^{-1} + L_2(m)^{-1}. \tag{7.9}$$

This last equation relates to the use of a symmetric V-mask where we want to test movements from the target value m_a of $m_a \pm \Delta$. Situations, of course, arise where we need to control m_a to limits of $m_a + \Delta_1$ and $m_a - \Delta_2$. We would do so by using a mask with limbs inclined at different angles α_1 and α_2 to the horizontal OP. Obvious modifications to the above show that for $\alpha_1 > \alpha_2$ (say) eqn (7.10) holds when

$$\omega \sin(\alpha_1 + \alpha_2) > d \sin(\alpha_1 - \alpha_2).$$

Consideration of the inequalities which lead to this conclusion indicate that we can expect (7.9) to hold to a reasonable level of approximation over a wide range of values for α_1 and α_2 which do not satisfy this last equation.

7.6 Wald sequential tests

For a Wald sequential test with horizontal decision boundaries 0 and h the sum

$$Z_n = y + \sum_{i=1}^{i=n} X_i \quad (0 \leq y \leq h)$$

is computed until the first value of $n = N$ arises, such that Z_N is ≤ 0 or $\geq h$, so that $0 < z_i < h$ for $i = 1$ to $i = N - 1$.

Let us take the case when X_i takes both negative and positive values and has f.f. $f(x; \theta)$. It has been shown (Wald 1944) that the probability that such a test will eventually cross one of its boundaries is 1. Write $p(y; \theta)$ for the probability that a test which starts distance y from the lower boundary will end on or below it. Thus, the probability that a test will terminate with $z_N \geq h$ is $1 - p(y; \theta)$. Let us use $N(y|z_N \leq 0; \theta)$ for the expected value of N conditional on $z_N \leq 0$ together with $N(y|z_N \geq h; \theta)$ for its expected value when $z_N \geq h$. If $N(y; \theta)$ is the unconditional average value of N

$$N(y; \theta) = N(y|z_N \leq 0; \theta)p(y; \theta) + N(y|z_N \geq h; \theta)[1 - p(y; \theta)].$$

$N(y; \theta)$ is called the average sample number of a Wald test. Evidently, assessing changes away from a target value by using plotted values of $S_1(j)$ is equivalent to a sequence of these tests with y equal to zero, and taking action on the first occasion that $S_1(j) \geq h$. The probability that there will be r members of this sequence before this occurs is evidently $p(0; \theta)^r$ so that the expected value of r is given by

$$\sum_{r=1}^{r=\infty} rp(0; \theta)^r[1 - p(0; \theta)] = p(0; \theta)/[1 - p(0; \theta)].$$

Thus, $L(\theta)$ the ARL of a single-sided scheme to keep θ on target is

$$L(\theta) = N(0|z_N \leq 0; \theta)\{p(0; \theta)/[1 - p(0; \theta)]\} + N(0|z_N \geq h; \theta)$$
$$= N(0; \theta)/[1 - p(0; \theta)]. \tag{7.10}$$

We therefore see that when we know the values of $p(0; \theta)$ and $N(0; \theta)$ of the Wald test which correspond to a particular cusum control chart, we can find $L(\theta)$ from this last equation. In view of this conclusion consider an identity due to Wald (1944) which we shall need to formulate expressions for $L(\theta)$.

7.7 Some assumptions

For convenience let us assume that X_i is a continuous variate. Consider the sum

$$z_n = y + \sum_{i=1}^{i=n} x_i$$

where y is fixed and x_i ($i = 1$ to n) are n independent values of variate X_i with f.f. $f(x; \theta)$. We need to establish some formal relationships which exist between the probabilities that a Wald sequential test will end in a particular way and a number of conditional expectations of Z_N and N. Let

- $g_n(z_n; y)$ and $G_n(z_n; y)$ respectively denote the f.f. and d.f. of Z_n;
- $E[\alpha(z_n); y]$ denote the expected value of a function $\alpha(z_n)$ of z_n and in the usual way use $E[\alpha(z_n); y|C]$ for its expected value subject to condition C;
- $P_h(y)$ be the probability that for all N, $Z_N \geq h$ whilst $P_0(y)$ is the corresponding probability that $Z_N \leq 0$.

As previously, $M(t)$ is the moment generating function of X_i. Assume that $M(t)$ is finite for all values of t as are its first two derivatives with respect to t. Let us further assume that these two derivatives can be obtained by differentiation within integrals. We have already seen in Chapter 3 that when X_i takes both positive and negative values with non-zero probabilities, the equation $M(t) = 1$ has a non-zero root τ so that $M(\tau) = 1$. Wald has shown that when these conditions are satisfied

$$P_h(y) + P_0(y) = 1.$$

In view of this result we shall write $P_0(y) = P(y)$ and $P_h(y) = 1 - P(y)$.

7.8 Wald's identity

Evidently $g_n(z_n; y)$ satisfies the equation

$$g_n(z_n; y) = \int_0^h g_{n-1}(z_{n-1}; y) f(z_n - z_{n-1}; \theta) \, dz_{n-1} \qquad (7.11)$$

so that

$$G_n(z_n; y) = \int_0^h G_{n-1}(z_{n-1}; y) f(z_n - z_{n-1}; \theta) \, dz_{n-1}. \tag{7.12}$$

Consider the sum

$$S_m = y + \sum_{i=1}^{i=m} X_i$$

where m is any positive integer and S_m is not subject to the restrictions of Z_n, namely that $0 \le Z_n \le h$. The d.f. $F(s_m; \theta)$ of S_m is that of the sum of m independent identically distributed variates with f.f. $f(x; \theta)$ so that

$$E(e^{tS_m}) = e^{ty}[M(t)]^m. \tag{7.13}$$

The unrestricted sums S_j can be separated into two mutually exclusive sets; those for which $0 \le S_j \le h$ for all $1 \le j \le m$, and those for which one or more values of S_j do not belong to this set. For the first set $N \ge m$, for the second j is equal to the lowest value of j at which S_j is either < 0 or $> h$. Evidently,

$$E(e^{tS_m}; y|N \le m) = E\left\{ \exp\left[t\left(Z_N + \sum_{i=N+1}^{i=m} X_i \right) \right]; y \mid N \le m \right\},$$

so that

$$E[e^{tZ_N} M(t)^{m-N}; y|N \le m]$$

$$= \left\{ \sum_{N=1}^{N=m} \int e^{tz_N} M(t)^{m-N} g_N(z_N; y) \, dz_N \right\} \Big/ p(m; y) \tag{7.14}$$

where $\int_{z_N}$ denotes integration over the ranges $-\infty$ to 0 and h to ∞ and z_N

$$p(m; y) = \sum_{N=1}^{N=m} \int_{z_N} g_N(z_N; y) dz_N.$$

In addition we have

$$E(e^{tS_m}; y|N > m) = \left[\int_0^h e^{tS_m} g_m(s_m; y) \, ds_m \right] \Big/ \left[\int_0^h g_m(s_m; y) \, ds_m \right]. \tag{7.15}$$

Taking eqns (7.14) and (7.15) together with (7.12) we find

$$E[e^{tZ_N} M(t)^{m-N}; y|N \le m] p(m; y) + [1-p(m; y)] \int_0^h e^{tS_m} g_m \, ds_m = e^{ty}[M(t)]^m. \tag{7.16}$$

Since by assumption $M(t)$ is finite for finite t and the limit as $m \to \infty$ of $p(m; y)$ is 1, it follows that for all finite t

$$\lim_{m \to \infty} E[e^{tZ_N} M(t)^{-N}; y | N \le m] = E\{e^{tZ_N} \cdot [M(t)]^{-N}; y\} = e^{ty} \qquad (7.17)$$

and that

$$E(e^{\tau Z_N}; y) = e^{\tau y}. \qquad (7.18)$$

Expressions which can be used to obtain values for $P(y)$ and the average sample number $E(N; y)$ of a particular Wald scheme can be derived from the expressions above, provided we can differentiate within integral signs and interchange orders of integration. Suppose we take eqn (7.17), differentiation with respect to t gives

$$E\{Z_N e^{tZ_N} \cdot [M(t)^{-N}; y\} - M'(t) E\{N e^{tZ_N} \cdot [M(t)]^{-N-1}; y\} = y e^{ty}$$

and taking $t = 0$ we have

$$E(Z_N; y) = M'(0) E(N; y) + y$$

and since $M'(0) = \theta$

$$E(Z_N; 0) = \theta E(N; 0). \qquad (7.19)$$

We can write eqn (7.17) in the form

$$P(y) E\{e^{tZ_N} \cdot [M(t)]^{-N}; y | Z_N \le 0\} + [1 - P(y)]$$

$$E\{e^{tZ_N} \cdot [M(t)]^{-N}; y | Z_N \ge h\} = e^{ty}. \qquad (7.20)$$

We can formulate integral equations which the conditional expectations in this and similar equations satisfy. To do so we need the following result, namely

$$G_n(z_n; y) = \int_0^h G_{n-1}(z_n; u) f(u - y; \theta) \, du. \qquad (7.21)$$

From eqn (7.12)

$$G(z_{n+1}; y) = \int_0^h G_n(z_n; y) f(z_{n+1} - z_n; \theta) \, dz_n,$$

so that if (7.21) is true

$$G(z_{n+1}; y) = \int_0^h \int_0^h G_{n-1}(z_n; u) f(u - y; \theta) f(z_{n+1} - z_n; \theta) \, du \, dz_n.$$

Reversing the order of integration gives

$$G_{n+1}(z_{n+1}; y) = \int_0^h G_n(z_{n+1}; u) f(u - y; \theta) \, du.$$

Since $G_1(z_1; y) = F(z_1 - y; \theta)$ it readily follows that

$$G_2(z_2; y) = \int_0^h G_1(z_2; u) f(u - y; \theta) \, du$$

and so (7.21) follows by induction.

We can write

$$P(y) E(e^{tZ} N; y | Z_N \leq 0) = \int_{-\infty}^0 e^{tz_1} f(z_1 - y; \theta) \, dz_1$$

$$+ \sum_{N=2}^{N=\infty} \int_{-\infty}^0 \int_0^h e^{tz} N g_{N-1}(z_N; u) f(u - y; \theta) \, du \, dz_N$$

so that

$$P(y) E(e^{tZ} N; y | Z_N \leq 0) = \int_{-\infty}^0 e^{tz_1} f(z_1 - y; \theta) \, dz_1$$

$$+ \int_0^h \sum_{N=2}^{N=\infty} \int_{-\infty}^0 e^{tz} N g_{N-1}(z_N; u) f(u - y; \theta) \, dz_N \, du$$

and

$$P(y) E(e^{tZ} N; y | Z_N \leq 0) = \int_{-\infty}^0 e^{tz_1} (f(z_1 - y; \theta) \, dz_1$$

$$+ \int_0^h P(u) E(e^{tz} N; u | z_N \leq 0) f(u - y; \theta) \, du. \tag{7.22}$$

When $t = 0$ we obtain

$$P(y) = F(-y; \theta) + \int P(u) f(u - y) \, du. \tag{7.23}$$

On the other hand, if we differentiate eqn (7.22) with respect to t and then put $t = 0$ we have

$$P(y) E(Z_N; y | Z_N \leq 0) = \int_{-\infty}^0 z_1 f(z_1 - y; \theta) \, dz_1$$

$$+ \int_0^h P(u) E(Z_N; u | Z_N \leq 0) f(u - y; \theta) \, du. \tag{7.24}$$

Using a similar procedure we also obtain

$$[1 - P(y)] E(Z_N; y | Z_N \geq h) = \int_h^\infty z_1 f(z_1 - y; \theta) \, dz_1$$

$$+ \int_0^h [1 - P(u)] E(Z_N; u | Z_N \geq h) f(u - y; \theta) \, du. \tag{7.25}$$

When we add these two equations together we find that

$$E(Z_N; y) = y + \theta - \int_0^h z_1 f(z_1 - y) \, dz_1 + \int_0^h E(Z_N; u) f(u - y) \, du. \quad (7.26)$$

The following result is obtained by using identical arguments, namely,

$$P(y)E(e^{tZ_N} \cdot M(t)^{-N}; y|Z_N \leq 0) = \left\{ \int_{-\infty}^0 e^{tz_1} f(z_1 - y; \theta) \, dz_1 \right.$$

$$\left. + \int_0^h P(u)E(e^{tz}N \cdot M(t)^{-N}; u|z_N \leq 0) f(u - y; \theta) \, du \right\} M(t)^{-1}. \quad (7.27)$$

A similar result obviously holds for $[1 - P(y)]E(e^{tZ_N} \cdot M(t)^{-N}; y|Z_N \geq h)$. Differentiating this last equation with respect to t and then putting $t = 0$ shows that

$$P(y)E(N; y|Z_N \leq 0) = P(y) + \int_0^h P(u)E(N; u|Z_N \leq 0) f(u - y; \theta) \, du$$

and

$$[1 - P(y)]E(N; y|Z_N \geq h)$$

$$= [1 - P(y)] + \int_0^h [1 - P(u)]E(N; u|Z_N \geq h) f(u - y; \theta) \, du.$$

Addition of these two equations then gives

$$E(N; y) = 1 + \int_0^h E(N; u) f(u - y; \theta) \, du. \quad (7.28)$$

7.9 Some useful relationships

Consider the run length R of a sequence of independent Wald tests of the kind just described, where the first test starts distance y from the lower boundary and subsequent tests start upon it. If $E(R; y)$ is the ARL of this sequence, then

$$E(R; y) = E(N; y) + P(y)E(R; 0),$$

so as we have already seen

$$E(R; 0) = E(N; 0)/[1 - P(0)].$$

Let us consider situations where it is not possible to obtain formal expressions for the evaluation of $P(y)$ and $E(N; y)$ which give their exact values. We shall then need to develop methods of quadrature to obtain them or devise expressions which give close approximations to them. In these circumstances there has been a tendency to resort to estimating $E(R; 0)$ by simulation, performing a number of computerised cumulative sum tests and obtaining their average value of R. We shall show that this is not the best or simplest way to approach the problem. With the computing facilities now available an easy way to compute the values of $P(0)$ and $E(N; 0)$ is to obtain the distribution function of Z_N using numerical methods of integration and the generating eqn (7.29) given below. This is particularly the case when reasonably approximate solutions to eqns like (7.23) and (7.28) indicate the values of h and θ needed to obtain specified run lengths. We shall see that we can devise such expressions which give very close upper and lower bounds for $P(y)$ and $E(N; y)$. Successive substitution of these into eqns (7.23) and (7.28) give values of $P(y)$ and $E(N; y)$ which quickly converge to their true values. We illustrate that a combination of these methods leads to simple expressions which can be used to determine parameter values for the design of optimal cumulative sum control schemes.

We begin with an illustration of the numerical computation of $P(0)$ and $E(N; 0)$, evidently

$$P(0) = \sum_{N=1}^{N=\infty} G_N(Z_N; 0)$$

and

$$E(N; 0) = \sum_{N=1}^{N=\infty} [G_N(Z_N; \infty) + G_N(Z_N; 0) - G_N(Z_N; h)] \cdot N$$

with

$$G_N(Z_N; 0) = \int_0^h F(z_N - z_{N-1}; \theta) g_{N-1}(z_{N-1}; 0) \mathrm{d}z_{N-1}. \tag{7.29}$$

Experience gained in calculating $G_N(Z_N; 0)$ indicates that the probability that N is $\geq 3E(N; Y)$ is very small. For most cumulative sum schemes used in practice the value of $E(N; 0)$ will not exceed 5. Table 7.1 illustrates that after a certain point the ratio of $G_{N-1}(Z_N; 0)$ to $G_N(Z_N; 0)$ approaches a constant value which we can use to terminate calculations with little effect on the value of $P(0)$. We accordingly find that the value of N required to obtain $P(0)$ and $E(N; 0)$ is not all that large. The computations required make little demand on the capabilities of contemporary computers.

Table 7.1 gives values $G_N(0; 0)$ and $G_N(\infty; 0) - G_N(h; 0)$ when X_i is a normal variate with unit standard deviation and mean θ equal to -0.50 and 1.00 when $h = 3$. Suppose we write λ_m^{-1} for the ratio

$$[G_{m-1}(\infty; 0) - G_{m-1}(h; 0)]/[G_m(\infty; 0) - G_m(h; 0)]$$

Table 7.1   Values of $G_N(0; 0)$ and $G_N(\infty; 0) - G_N(h; 0)$ for $h = 3$, when X_i is $N(\theta; 1)$ with $\theta = -0.50$ and $\theta = 1.00$.

	$\theta = -0.50$		$\theta = 1.00$	
N	$G_N(0)$	$G_N(\infty) - G_N(h)$	$G_N(0)$	$G_N(\infty) - G_N(h)$
1	0.691463	0.000233	0.158655	0.002275
2	0.141055	0.002330	0.026731	0.217719
3	0.061152	0.003219	0.008301	0.249747
4	0.033350	0.002817	0.003166	0.156751
5	0.020201	0.002085	0.001329	0.081036
6	0.012872	0.001451	0.000584	0.038998
7	0.008398	0.000984	0.000263	0.018221
8	0.005541	0.000660	0.000119	0.008412
9	0.003673	0.000441	0.000054	0.003863
10	0.002440	0.000294	0.000025	0.001517
11	0.001622	0.000196	0.000012	0.000809
12	0.001079	0.000130	0.000005	0.000372
13	0.000718	0.000087	0.000002	0.000171
14	0.000478	0.000058	0.000001	0.000078
15	0.000318	0.000038		0.000027
16	0.000216	0.000026		0.000016
17	0.000140	0.000017		0.000007
18	0.000094	0.000011		
19	0.000062	0.000008		
20	0.000041	0.000005		

then when this ratio settles down to a reasonably constant value we can write

$$P(0) \cong \sum_{N=1}^{N=m} G_N(0; 0) + G_m(0; 0)\lambda_m/(1 - \lambda_m). \tag{7.30}$$

From Table 7.1 we see that for $\theta = -0.50$, $\lambda_{10} = 0.6643$. If we use the approximation (7.30) we obtain $P(0) = 0.98497$. The true value of $P(0)$ is 0.984982. The corresponding truncated expression for $E(N; 0)$ is

$$E(N; 0) = \sum_{N=1}^{N=m} [G_N(0; 0) + G_N(\infty; 0) - G_N(h; 0)] \cdot N$$

$$+ [G_m(0; 0) + G_m(\infty; 0) - G_m(h; 0)]$$

$$\cdot [(m + 1) - m\lambda_m]\lambda_m/(1 - \lambda_m)^2. \tag{7.31}$$

When $m = 10$ this equation gives a value for $E(N; 0)$ of 1.5758 in the above case. Truncated values for $P(0)$ and $E(N; 0)$ therefore give $E(R; 0) = 104.8$. Its true value is 104.5.

7.10 The run length distribution

If $E(R; 0)$ is to be obtained by simulating cumulative sums we need to establish some basic features of the distribution of R. This is certainly so when testing is in control and $E(R; 0)$ is therefore large. We can do this by using the equation for $\mathbf{P(R; 0)}$ the probability that R takes a specific value namely, $r = R - 1$,

$$\mathbf{P(R; 0)} = G_R(\infty; 0) - G_R(h; 0) + \sum_{r=1}^{r=R-1} G_r(0; 0)\mathbf{P(R - r; 0)}. \tag{7.32}$$

To do so, let us write $M_R(t)$ for the moment generating function of R, $M_N(t)$ for that of N, with $M_N(t|Z_N \le 0)$ and $M_N(t|Z_N \ge h)$ for the m.g.fs. of N conditional on $Z_N \le 0$ and $Z_N \ge h$. When these m.g.fs. exist

$$M_R(t) = \sum_{R=1}^{R=\infty} \mathbf{P(R; 0)}e^{Rt} \tag{7.33}$$

and

$$M_N(t) = \sum_{N=1}^{N=\infty} [G_N(\infty; 0) - G_N(h; 0) + G_N(0; 0)] \cdot e^{Nt}.$$

Use of eqns (7.32) and (7.33) with an obvious re-ordering of terms in the resulting double summation then gives

$$M_R(t) = [1 - P(0)]M_N(t|Z_N \ge h)/[1 - P(0)M_N(t|Z_N \le 0)]$$
$$= 1 + [M_N(t) - 1]/[1 - P(0)M_N(t|Z_N \le 0)]. \tag{7.34}$$

Assuming the first two differentials with respect to t exist for the m.g.fs. in this equation we obtain that the variance $V(R)$ of R is

$$V(R) = [V(N) + E(R; 0)P(0)\{2E(N|Z_N \le 0)]/[1 - P(0)] - E(R; 0)\}.$$

Since at AQL $P(0)$ is very nearly 1 and

$$E(N|Z_N \le 0)P(0)/[1 - P(0)] = E(R; 0) - E(N|Z_N \ge h) \cong E(R; 0)$$

we have

$$V(R) \cong \{V(N)/[1 - P(0)]\} + E^2(R; 0). \tag{7.35}$$

We therefore have the important result that the standard deviation of R exceeds $E(R; 0)$. We can use this result to estimate the number of simulations which are required to obtain a value for $E(R; 0)$ to a given degree of accuracy. Suppose we require the parameters of a control scheme with $L(0) = E(R; 0)$ in the region of 500 and R_j is the run obtained for the jth simulation. If $\mathbf{S}$ is the number of values of X_i used in a single simulation, then since $\mathbf{S}$ will be large we can assume that

$$\bar{R} = \sum_{j=1}^{j=\mathbf{S}} R_j / \mathbf{S}$$

is normally distributed with mean $E(R; 0)$ and standard deviation $\sigma(\bar{R}) > E(R; 0)/\sqrt{\mathbf{S}}$. If we require the probability that the mean value of $\mathbf{S}$ simulations will lie in the range 0.95 $E(R; 0)$ to 1.05 $E(R; 0)$ to be 0.998 (say); that is,

$$p[0.95E(R; 0) \le \bar{R} \le 1.05E(R; 0)] = 0.998,$$

then since $E(R; 0) = 500$ and $\sigma(\bar{R}) > 8.09$ the value of $\mathbf{S}$ is greater than 3820. We therefore find that to obtain an accuracy of ± 5 per cent in the estimation of $E(R; 0)$ when it is 500, we would need some two million random numbers to carry through the simulation. Evidently if, as indicated in Chapter 1, we use a pseudo random number generator to do so, it will have to be chosen with very great care. The generator will obviously need to have a cycle well in excess of $2 \cdot 10^6$. The formation of such a large number of values with pseudo randomness to obtain values of $E(R; 0)$ fills one with unease with regard to the validity of the answers such a method will give. One way to possibly avoid this problem might be to obtain values of $E(N; 0)$ and $P(0)$ for the constituent Wald test using methods of simulation. We could then use eqn (7.10) to obtain $E(R; 0)$. It has to be said that this method would require the value of $P(0)$ to be very accurate indeed so that the degree of simulation needed is also likely to be considerable.

We can show that for a sequence of Wald tests, where the first test starts distance y from the lower boundary and subsequent ones start on it, has run length $E(R; y)$ which satisfies the equation

$$E(R; y) = 1 + E(R; 0)F(-y; \theta) + \int_0^h E(R; u)f(u - y; \theta)\,\mathrm{d}u. \qquad (7.36)$$

Computational difficulties can arise, however, with use of this equation to find values of $E(R; 0)$ directly. It is accordingly better to obtain $P(0)$ and $E(N; 0)$ from the equations already given and use eqn (7.10).

Notation

It will now be convenient to use the following notation,

- $E[\xi(Z_N); y]$ for the unconditional expected value of the function $\xi(Z_N)$ of Z_N;
- $E_0[\xi(Z_N); y]$ for the expected value of $\xi(Z_N)$ conditional on $Z_N \le 0$;
- $E_h[\xi(Z_N); y]$ for the expected value of $\xi(Z_N)$ conditional on $Z_N \ge h$;

In this notation Wald's identity is obviously

$$P(y)E_0(\mathrm{e}^{\tau Z_N}; y) + [1 - P(0)]E_h(\mathrm{e}^{\tau Z_N}; y) = \mathrm{e}^{\tau y}$$

whilst eqn (7.22) can be written as

$$P(y)E_0(e^{\tau Z_N}; y) = \int_{-\infty}^{0} e^{\tau u} f(u - y; \theta)\, du$$

$$+ \int_{0}^{h} P(u)E_0(e^{\tau Z_N}; u) f(u - y; \theta)\, du.$$

7.11 An expression for $P(0)$

The methods we now describe consist of formulating 'approximating' equations for $P(0)$ and $E(N; 0)$. We shall see that these yield values of $L(0)$ which are so close to the true values of $L(0)$ that they can be used to design control schemes with estimated run lengths which are sufficiently accurate for practical purposes. They can also be used to write subroutines for fast converging search programmes to obtain additional accuracy for $L(0)$ when this may be required. In view of this last remark it is perhaps appropriate to emphasize that the accuracy of the expressions given is such that for most practical applications this procedure is unnecessary. We shall see that the latter merely provide a basis for theoretical comparisons rather than the needs of schemes to be used for routine quality control procedures.

It will be convenient to write the m.g.f. of Z_N given y as $M(t; y)$ and those of Z_N conditional on $Z_N \leq 0$ and $Z_N \geq h$ as $M_0(t; y)$ and $M_h(t; y)$. Further, we will write $\mu(t; y)$, $\mu_0(t; y)$ and $\mu_h(t; y)$ for the corresponding m.g.fs. of Z_N when $N = 1$. In this notation $\mu(t; y) = e^{\tau y}$ and eqn (7.22) is

$$P(y)M_0(\tau; y) + [1 - P(y)]M_h(\tau; y) = e^{\tau y}$$

whilst

$$\mu_0(\tau; y) = \int_{-\infty}^{0} e^{\tau z_1} f(z_1 - y; \theta)\, dz_1 / F(-y; \theta)$$

and

$$\mu_h(\tau; y) = \int_{h}^{\infty} e^{\tau z_1} f(z_1 - y; \theta)\, dz_1 / [1 - F(h - y; \theta)].$$

From the mean value theorem it follows from eq. (7.22) that

$$P(y)\{M_0(\tau; y) - M_0[\tau; \lambda_0(y)]\} = \{\mu_0(\tau; y) - \mu_0[\tau; \lambda_0(y)]\} F(-y; \theta) \quad (7.37)$$

where $\lambda_0(y)$ is a function of y such that $0 \leq \lambda_0(y) \leq h$. In similar fashion, we have

$$[1 - P(y)]\{M_h(\tau; y) - M_h[\tau; \lambda_h(y)]\} \quad (7.38)$$
$$= \{\mu_h(\tau; y) - \mu_h[\tau; \lambda_h(y)]\}[1 - F(h - y; \theta)].$$

If we add these two equations together, and for brevity write $\lambda_0(y) = \lambda_0$ and similarly $\lambda_h(y) = \lambda_h$ we get

$$[1 - P(y)] = \big(\{M_0(\tau; y) - \mu_0(\tau; \lambda_0)\}F(-y; \theta) + \{M_h(\tau; y) - \mu_h[\tau; \lambda_h]\}$$

$$[1 - F(h - y; \theta)]\big)\big/[M_h(\tau; \lambda_h) - M_0(\tau; \lambda_0)]. \qquad (7.39)$$

Now let y_M be the value of y at which $M_0(\tau; y)$ takes its maximum value. Then from (7.22) and (7.23) it follows that,

$$P(y_M)M_0(\tau; y_M) \leq \mu_0(\tau; y_M)F(-y_M; \theta) + M_0(\tau; y_M)[P(y_M) - F(-y_M; \theta)]$$

and therefore

$$M_0(\tau; y_M) \leq \mu_0(\tau; y_M) \leq \max_{0 \leq y \leq h} \mu_0(\tau; y) \qquad (7.40)$$

Similarly, if y_m is the value of y at which $M_0(\tau; y)$ takes its minimum value we have

$$M_0(\tau; y_m) \geq \min_{0 \leq y \leq h} \mu_0(\tau; y). \qquad (7.41)$$

We can therefore find a value of y between 0 and h so that for $0 \leq u \leq h$

$$\mu_0(\tau; y) = M_0(\tau; u).$$

Denote this value by y_0 when $y = 0$ so that $\mu_0(\tau; y_0) = M_0[(\tau; \lambda_0(0)]$. Using similar reasoning we can find y_h such that $\mu_h(\tau; y_h) = M_h[\tau; \lambda_h(0)]$, we then obtain the result

$$[1 - P(0)] = \big\{1 - \mu_0(\tau; 0) + [\mu_0(\tau; 0) - \mu_0(\tau; y_0)][1 - F(0; \theta)]$$

$$+ [\mu_h(\tau; y_h) - \mu_h(\tau; 0)[1 - F(h; \theta)]\big\}$$

$$\big/[\mu_h(\tau; y_h) - \mu_0(\tau; y_0)]. \qquad (7.42)$$

7.12 Average sample number $E(N; 0)$

We can derive a corresponding expression for $E(N; 0)$ using eqns (7.24) to (7.26). Write $\beta_0(y)$ and $\beta_h(y)$ for the expectations of Z_1 conditional on $Z_1 \leq 0$ and $Z_1 \geq h$ so that

$$\beta_0(y) = \int_{-\infty}^{0} z_1 f(z_1 - y; \theta)dz_1 / F(-y; \theta)$$

and

$$\beta_h(y) = \int_{h}^{\infty} z_1 f(z_1 - y; \theta)dz_1 / [1 - F(h - y; \theta)].$$

Application of the mean value theorem leads to

$$P(y)\{E_0(Z_N; y) - E_0[Z_N; \gamma_0(y)]\} = \{\beta_0(y) - E_0[Z_N; \gamma_0(y)]\}F(-y; \theta)$$

$$(7.43)$$

and

$$[1 - P(y)]\{E_h(Z_N; y) - E_h(Z_N; \gamma h(y))\} = \{\beta_h(y) - E_h[Z_N; \gamma_h(y)]\}$$
$$\times [1 - F(h - y; \theta)], \qquad (7.44)$$

with $0 \le \gamma_0(y) \le h$ and $0 \le \gamma_h(y) \le h$. As previously values Y_0 and Y_h can be found such that

$$\beta_0(Y_0) = E_0[Z_N; \gamma_0(0)] \quad \text{and} \quad \beta_h(Y_h) = E_h[Z_N; \gamma h(0)].$$

Substitution of these into eqns (7.43) and (7.44) and using (7.19) gives

$$\theta E(N; 0) = \beta_0(0) + [\beta_0(Y_0) - \beta_0(0)][1 - F(0; \theta)]$$
$$+ [\beta_h(0) - \beta_h(Y_h)][1 - F(h; \theta)]$$
$$+ [1 - P(0)][\beta_h(Y_h) - \beta_0(Y_0)]. \qquad (7.45)$$

Values of $L(0)$ can be computed from these expressions when y_0, y_h, Y_0, and Y_h are known. Simple analytic consideration of eqns (7.42) and (7.45) together with some exploratory computations of $P(0)$ and $E(N; 0)$ obtained by quadrature, show that they can be used to derive

(i) easily programmed expressions which give very close estimates of $L(0)$ for a wide range of values of h and θ;

(ii) simple approximate expressions which do not require sophisticated computing equipment to obtain values of $L(0)$.

As we have already implied, the need to make these two distinctions arises from the differing demands of theory and practice. Expressions with the accuracy of (i) are necessary to confirm the existence of optimal cusum schemes. Those of (ii) relate to practical circumstances where the availability of good approximating equations for $L(0)$ is sufficient.

7.13 Determination of y_0, y_h, Y_0, and Y_h

Since

$$E_0(e^{\tau Z_N}; y) = \int_{-\infty}^{0} e^{\tau x} \sum_{N=1}^{N=\infty} g_N(x; y)\, dx \bigg/ \sum_{N=1}^{N=\infty} G_N(0; y),$$

it follows from eqn (7.37) that

$$\int_{-\infty}^{0} e^{\tau x} \sum_{N=1}^{N=\infty} g_N(x; y) \, dx = \int_{-\infty}^{0} e^{\tau x} f(x - y; \theta) \, dx + M_0[\tau; \lambda_0(y)]$$

$$\cdot \int_{0}^{h} P(u) f(u - y; \theta) \, du.$$

In virtue of (7.18) we have in particular that,

$$\mu_0(\tau; y_0) = \left\{ \int_{-\infty}^{0} e^{\tau x} \left[\sum_{N=1}^{N=\infty} g_N(x; 0) - f(x; \theta) \right] dx \right\} \Big/ [P(0) - F(0; \theta)].$$

$$(7.46)$$

We can use this last equation to determine y_0 when values of $g_N(x; 0)$ are known. Since y_h is functionally related to y_0 its value can be obtained from that of y_0. Thus by definition and Wald's identity we get

$$[1 - P(y)]E_h(e^{\tau Z_N}; y) = \int_{h}^{\infty} e^{\tau x} \sum_{N=1}^{N=\infty} g_N(x; y) dx$$

$$= e^{\tau y} - \int_{-\infty}^{0} e^{\tau x} \sum_{N=1}^{N=\infty} g_N(x; y) dx.$$

From eqn (7.46) and its equivalent to (7.22) for $[1 - P(y)]E_h(e^{\tau Z_N}; y)$ we find that

$$\mu_h(\tau; y_h) = \left\{ \int_{0}^{h} e^{\tau x} f(x; \theta) \, dx - \mu_0(\tau; y_0)[P(0) - F(0; \theta)] \right\}$$

$$\Big/ [F(h; \theta) - P(0)].$$

$$(7.47)$$

Corresponding expressions can be derived for Y_0 and Y_h, namely,

$$\beta_0(Y_0) = \left\{ \int_{-\infty}^{0} x \left[\sum_{N=1}^{N=\infty} g_N(x; y) - f(x; \theta) \right] dx \right\} \Big/ [P(0) - F(0; \theta)].$$

$$(7.48)$$

and

$$\beta_h(Y_h) = \left\{ \int_{0}^{h} x f(x; \theta) \, dx - \beta_0(Y_0)[P(0) - F(0; \theta)] \right\} \Big/ [F(h; \theta) - P(0)].$$

$$(7.49)$$

7.14 Iteration for values of $P(y)$ and $E(N; y)$

For many frequency functions $f(x; \theta)$ which arise in practice it is difficult to formulate expressions which give exact values for $P(y)$ and $E(N; y)$. This is the case for the equations

$$P(y) = F(-y; \theta) + \int_0^h P(u) f(u - y; \theta) \, du$$

and

$$E(N; y) = 1 + \int_0^h E(N; u) f(u - y; \theta) \, du$$

when X is a normal variate. From a practical point of view, however, they are not as intractable as may at first sight appear. We find that useful procedures can be devised using the above equations in a comparatively crude fashion in conjunction with currently available computing facilities. We can obtain values for both $P(y)$ and $E(N; y)$ iteratively. Evidently $E(N; y) > 1$, and so it follows that

$$E(N; y) > 1 + \int_0^h f(u - y; \theta) \, du.$$

If we write

$$E_1^{(1)}(N; y) = 1 + F(h - y; \theta) - F(-y; \theta),$$

then

$$E(N; y) \geq 1 + \int_0^h [1 + F(h - u; \theta) - F(-u; \theta)] f(u - y; \theta) \, du$$

$$= E_1^{(1)}(N; y) + \int_0^h [F(h - u; \theta) - F(-u; \theta)] f(u - y; \theta) \, du.$$

If we use $E_1^{(2)}(N; y)$ for this second approximation to $E(N; y)$ then

$$E_1^{(2)}(N; y) = 1 + \int_0^h E_1^{(1)}(N; u) f(u - y; \theta) \, du. \tag{7.50}$$

Evidently $E_1^{(1)}(N; y)$ is a lower bound for $E(N; y)$ for all $0 \leq y \leq h$ as is $E_1^{(2)}(N; y)$. Furthermore if, we repeat the successive substitution into the above equation so that

$$E_1^{(r)}(N; y) = 1 + \int_0^h E_1^{(r-1)}(N; u) f(u - y; \theta) \, du$$

then $E_1^{(r-1)}(N; y) \leq E_1^{(r)}(N; y)$. It can easily be shown (Kemp 1970) that $E_1^{(r)}(N; y)$ converges from below to $E(N; y)$. Sequences of upper and lower bounds for $P(y)$ will be generated for this probability, using the equation

$$P^{(r)}(y) = F(-y; \theta) + \int_0^h P^{(r-1)}(u) f(u - y; \theta) \, du. \tag{7.51}$$

Convergent sequences $P_u^{(r)}(y)$ and $P_1^{(r)}(y)$ for $P(y)$ can be formulated when $P_u^{(1)}(y) > P(Y)$ and $P_1^{(1)}(y) < P(y)$. Evidently we do not need to search for close approximations to $P(y)$ and $E(N; y)$ in order to use this method. We could for example start with $P_u(y) = 1$ and $P_1(y) = F(-y; \theta)$. However, we now see that very close initial upper and lower bounds for $P(y)$ and $E(N; y)$ can be formulated which give values for them which

- are sufficiently accurate for many practical purposes;
- assure fast convergence to the true values of $P(y)$ and $E(N; y)$ when the need arises.

7.15 Upper and lower bounds for $P(y)$ and $E(N; y)$

Consider expressions for $P(y)$ and take eqn (7.22), with their use Kemp (1970) demonstrated that we can derive upper and lower limits for $P(y)$ which are close to it. These can be used to form sequences $P_u^{(r)}(y)$ and $P_1^{(r)}(y)$ which converge from above and below to $P(y)$.

Let $y_{\max}$ and $y_{\min}$ be the values of y for which $E_0(e^{\tau Z_N}; y)$ takes its maximum and minimum values. From (7.22) and (7.23) we have that

$$P(y) E_0(e^{\tau Z_N}; y) \geq \int_{-\infty}^0 e^{\tau u} f(u - y; \theta) \, du$$

$$+ \int_0^h \int_{-\infty}^0 e^{\tau \omega} f(\omega - u; \theta) f(u - y; \theta) \, d\omega \, du$$

$$+ E_0(e^{\tau Z_N}; y_{\min})$$

$$\cdot \int_0^h \int_0^h P(\omega) f(\omega - u; \theta) f(u - y; \theta) \, d\omega \, du$$

and

$$P(y) - F(-y; \theta) - \int_0^h F(-u; \theta) f(u - y; \theta) \, du$$

$$= \int_0^h \int_0^h P(\omega) f(\omega - u; \theta) f(u - y; \theta) \, d\omega \, du.$$

It follows from these two equations that $E_0(e^{\tau Z_N}; y)$ is greater than the minimum value of the ratio

$$\psi_0(y; \theta) = \frac{\mu_0(\tau; y)F(-y; \theta) + \int_0^h \mu_0(\tau; u)F(-u; \theta)f(u - y; \theta)\,du}{F(-y; \theta) + \int_0^h F(-u; \theta)f(u - y; \theta)\,du}.$$

$$(7.52)$$

Denote the value of y at which this occurs by y_m, then $E_0(e^{\tau Z_N}; y) \geq \psi_0(y_m; \theta)$. In similar, fashion, if y_M is the maximum value of this ratio then $E_0(e^{\tau Z_N}; y) \leq \psi_0(y_M; \theta)$. If for $[1 - P(y)]E_h(e^{\tau Z_N}; y)$ we go through the same procedure, we obtain that $E_h(e^{\tau Z_N}; y) \geq \psi_h(y_m'; \theta)$ and $E_h(e^{\tau Z_N}; y) \leq \psi_h(y_M'; \theta)$ where

$$\psi_h(y; \theta) = \frac{\mu_h(\tau; y)F(h - y; \theta) + \int_0^h \mu_h(\tau; u)F(h - u; \theta)f(u - y; \theta)\,du}{F(h - y; \theta) + \int_0^h F(h - u; \theta)f(u - y; \theta)\,du}.$$

whilst y_m' and y_M' are the values of y for which the above ratio takes its minimum and maximum values.

If we now return to Wald's identity then clearly,

$$P(y)\psi_0(y_m; \theta) + [1 - P(y)]\psi_h(y_m'; \theta) \leq e^{\tau y},$$

so that

$$P(y) \geq [\psi_h(y_m'; \theta) - e^{\tau y}]/[\psi_h(y_m'; \theta) - \psi_0(y_m; \theta)] = P_l^{(1)}(y) \qquad (7.53)$$

and

$$P(y) \leq [\psi_h(y_M'; \theta) - e^{\tau y}]/[\psi_h(y_M'; \theta) - \psi_0(y_M; \theta)] = P_u^{(1)}(y). \qquad (7.54)$$

Expressions for $P_l^{(2)}(y)$ and $P_u^{(2)}(y)$ are obtained by substituting (7.53) and (7.54) into eqn (7.51). We obtain a weighted average of these two by using the following approximation argument. Let us write $P(y) = P_u^{(1)}(y)W_u(y)$ together with $P(y) = P_l^{(1)}(y)W_1(y)$ then from eqn (7.23) and the mean value theorem,

$$P(y) = F(-y; \theta) + \int_0^h W_u(\omega)P_u^{(1)}(\omega)f(\omega - y; \theta)\,d\omega$$

$$= F(-y; \theta) + W_u(\alpha h; y)\int_0^h P_u^{(1)}(\omega)f(\omega - y; \theta)\,d\omega$$

and

$$P(y) = F(-y; \theta) + W_1(\beta h; y) \int_0^h P_1^{(1)}(\omega) f(\omega - y; \theta) \, d\omega$$

where α and β lie between 0 and 1. Adding these two equations gives

$$2P(y) = F(-y; \theta)\{2 - [W_u(\alpha h; y) + W_1(\beta h; y)]\}$$

$$+ W_u(\alpha h; y) P_u^{(2)}(y) + W_1(\beta h; y) P_1^{(2)}(y). \tag{7.55}$$

Put $W_u(\alpha h; y) + W_1(\beta h; y) = 2 + \Delta(y)$ and consider the case where $\Delta(y)$ is small enough for $\Delta^2(y)$ to be ignored. Write $w_u(y)$ for the ratio

$$W_u(\alpha h; y)/[W_u(\alpha h; y) + W_1(\beta h; y)]$$

$$= [P_1^{(2)}(y) - F(-y; \theta)]/[P_u^{(2)}(y) + P_1^{(2)}(y) - 2F(-y; \theta)]$$

and $w_1(y) = 1 - w_u(y)$, then

$$w_u(y) P_u^{(2)}(y) + w_1(y) P_1^{(2)}(y) \cong F(-y; \theta) + [P(y) - F(-y; \theta)] \cdot [1 - \Delta(y)/2],$$

so that

$$P_w(y) = w_u(y) P_u^{(2)}(y) + w_1(y) P_1^{(2)}(y) \cong P(y).$$

We can obviously use the above reasoning to obtain the more general result that after r iterations,

$$P_w(y) = w_u(y) P_u^{(r)}(y) + w_1(y) P_1^{(r)}(y) \cong P(y).$$

It follows from eqns (7.51) and (7.55) that $P_w(0)$ should very closely approximate $P(0)$. In the example we give shortly for X being a normal variate we shall see that this is indeed the case.

A similar procedure using corresponding equations for $E(N; y)$ leads to the same conclusion with regard to the weighted average $E_w(N; 0)$ of $E_u(N; 0)$ and $E_1(N; 0)$. Thus, write Y_M and Y_m for the values of y at which the ratio

$$\xi_0(y; \theta) = \frac{\beta_0(y) F(-y; \theta) + \int_0^h \beta_0(u) F(-u; \theta) f(u - y; \theta) \, du}{F(-y; \theta) + \int_0^h F(-u; \theta) f(u - y; \theta) \, du}$$

takes its maximum and minimum values, and Y'_M and Y'_m for the corresponding values of y which maximize and minimize

$$\xi_h(y; \theta) = \frac{\beta_h(y) F(h - y; \theta) + \int_0^h \beta h(u) F(h - u; \theta) f(u - y; \theta) \, du}{F(h - y; \theta) + \int_0^h F(h - u; \theta) f(u - y; \theta) \, du}.$$

When $P_u(y)$ and $P_1(y)$ are upper and lower bounds for $P(y)$,

$$E(N; 0) \geq \{P_1(0)\xi_0(Y_M; \theta) + [1 - P_1(0)]\xi_h(Y_M; \theta)\}/E(x; \theta) = E_1(N; 0) \tag{7.56}$$

and

$$E(N; 0) \leq \{P_u(0)\xi_0(Y_m; \theta) + [1 - P_u(0)]\xi_h(Y_m; \theta)\}/E(x; \theta) = E_u(N; 0). \tag{7.57}$$

It follows that

$$E_w(N; 0) = w_u'(0)E_u(N; 0) + [1 - w_u'(0)]E_1(N; 0) \tag{7.58}$$

where

$$w_u'(0) = [E_1(N; 0) - 1]/[E_u(N; 0) + E_1(N; 0) - 2],$$

and

$$E_w(R; 0) = E_w(N; 0)/[1 - P_w(0)]. \tag{7.59}$$

7.16 The case of the normal distribution

We can illustrate the accuracy of eqn (7.59) by considering the situation when X is normally distributed with mean θ and unit standard deviation. Let us take θ to be negative. We have

$$\psi_0(y; \theta) = \frac{e^{-2\theta y}\Phi(\theta - y) + \int_0^h e^{-2\theta u}\Phi(\theta - u)\phi(u - y - \theta)\,du}{\Phi(-\theta - y) + \int_0^h \Phi(-\theta - u)\phi(u - y - \theta)\,du} \tag{7.60}$$

and

$$\psi_h(y; \theta) = \frac{e^{-2\theta y}\Phi(y - h - \theta) + \int_0^h e^{-2\theta u}\Phi(u - h - \theta)\phi(u - y - \theta)\,du}{\Phi(\theta - h + y) + \int_0^h \Phi(\theta - h + u)\phi(u - y - \theta)\,du} \tag{7.61}$$

with

$$
\xi_0(y; \theta) = \left\{ (y + \theta)\Phi(-y - \theta) - \phi(-y - \theta) \right.
$$

$$
\left. + \int_0^h (u + \theta)[\Phi(-u - \theta) - \phi(-u - \theta)]\phi(u - y - \theta)\, du \right\}
$$

$$
\left/ \left\{ \Phi(-y - \theta) + \int_0^h \Phi(-u - \theta)\phi(u - y - \theta)\, du \right\} \right. \tag{7.62}
$$

and

$$
\xi_h(y; \theta) = \left\{ (y + \theta)\Phi(y - h + \theta) + \phi(y - h + \theta) \right.
$$

$$
\left. + \int_0^h (u + \theta)[\Phi(u - h + \theta) + \phi(u - h + \theta)]\phi(u - y - \theta)\, du \right\}
$$

$$
\left/ \left\{ \Phi(\theta - h + y) + \int_0^h \Phi(\theta - h + u)\phi(u - y - \theta)\, du \right\} \right.
$$

$$
\tag{7.63}
$$

If we take the ratio $e^{-2\theta y}\Phi(\theta - y)/\Phi(-\theta - y)$ it is not difficult to show (Kemp 1967) that when θ is negative this ratio increases as y increases, it then follows that the ratios (7.60) and (7.61) take their minimum and maximum values at $y = 0$ and $y = h$ when θ is negative. For such θ we therefore have

$$
P_1^{(1)}(y) = [\psi_h(0; \theta) - e^{-2\theta y}]/[\psi_h(0; \theta) - \psi_0(0; \theta)]. \tag{7.64}
$$

From eqns (7.61) and (7.62) we find that $\psi_h(h; \theta) = e^{-2\theta h}/\psi_0(0; \theta)$ and the ratio $\psi_h(0; \theta) = e^{-2\theta h}/\psi_0(h; \theta)$ so that

$$
P_u^{(1)}(y) = [\psi_h(h\theta) - e^{-2\theta y}]/[\psi_h(h; \theta) - \psi_0(h; \theta)]
$$

$$
= \psi_h(0; \theta)[1 - \psi_0(0; \theta)e^{-2\theta(y-h)}]/[\psi_h(0; \theta) - \psi_0(0; \theta)].
$$

When we substitute these expressions into eqn (7.51) it generates values for $P_1^{(r)}(y)$ and $P_u^{(r)}(y)$ which converge to $P(y)$. We can find these by numerical methods when r is greater than 2. For $r = 2$ we can derive formal expressions for them. To calculate their values we need tables of the bivariate normal distribution function with correlation coefficient $1/\sqrt{2}$.

If we write

$$
F(x_1, x_2) = (\pi\sqrt{2})^{-1} \int_{-\infty}^{x_1} \int_{-\infty}^{x_2} \exp[-(x^2 - xy\sqrt{2} + y^2)]\, dx\, dy
$$

and

$$\Omega(u, y|\theta) = F\{[(u - 2\theta)/\sqrt{2}], (h - y - \theta)\} - F\{[(u - 2\theta)/\sqrt{2}], (-y - \theta)\}.$$

For convenience let us write

$$\mu = [\Phi(\theta) + \Omega(0, 0| - \theta)]/[\Phi(-\theta) + \Omega(0, 0|\theta)]$$

and

$$\xi = [\Phi(-\theta) - \Omega(h, 0| - \theta)]/[\Phi(\theta) - \Omega(h, 0|\theta)],$$

in this notation we find

$$P_u^{(2)}(y) = \left\{ \xi\Omega(h - y, y|\theta) - \mu\Omega(-y; y|\theta) - \xi\mu e^{-2\theta(h-y)} \right.$$
$$\left. \cdot [\Omega(h - y, y| - \theta) - \Omega(-y, y| - \theta)] \right\} \Big/ (\xi - \mu) \qquad (7.65)$$

and

$$P_1^{(2)}(y) = \left\{ \xi\Omega(h - y, y|\theta) - \mu\Omega(-y; y|\theta) - e^{-2\theta y} \right.$$
$$\left. \cdot [\Omega(h - y, y| - \theta) - \Omega(-y; y| - \theta)] \right\} \Big/ (\xi - \mu). \qquad (7.66)$$

Corresponding expressions can be derived for $E_1^{(2)}(N; y)$ and $E_u^{(2)}(N; y)$.

The weighted averages $P_w(0)$ and $E_w(0)$ given by these equations give very close approximations to $P(0)$ and $E(N; 0)$ and hence to $E(R; 0)$. A comparison between them and the true values is shown in Table 7.2 and Table 7.3 when $h = 3$. From these it is clear that these expressions can be used for most practical purposes.

Table 7.2   Upper and lower bounds for $P(0)$ and $E(N; 0)$ when X is $N(\theta; 1)$ and $h = 3$

θ	$P_1^{(2)}(0)$	$P_u^{(2)}(0)$	$E_1^{(2)}(N; 0)$	$E_u^{(2)}(N; 0)$
−0.40	0.9703	0.9773	1.74	2.13
−0.60	0.9904	0.9932	1.48	1.70
−0.80	0.9974	0.9981	1.31	1.44

Table 7.3   Weighted values $P_w^{(2)}(0)$, $E_w^{(2)}(N; 0)$, $E_w^{(2)}(R; 0)$ and the exact values of $P(0)$, $E(N; 0)$ and $E(R; 0)$

θ	$P_w^{(2)}(0)$	$E_w^{(2)}(N; 0)$	$E_w^{(2)}(R; 0)$	$P(0)$	$E(N; 0)$	$E(R; 0)$
−0.40	0.9783	1.84	72	0.9753	1.95	74
−0.60	0.9920	1.57	196	0.9919	1.61	198
−0.80	0.9977	1.36	591	0.9976	1.386	585

7.17 Simplified expressions for $E(R; 0)$

If we know the values of y_0, y_h, Y_0 and Y_h or approximations to them, we can use eqns (7.42) and (7.45) to estimate $P(0)$ and $E(N; 0)$. As we have already indicated, simpler if slightly more approximate expressions for $L(0) = E(R; 0)$ can be obtained for specific distributions of X when only dominant terms of these two equations are retained. The resulting formulae are easy to use and give values for $L(0)$ which are again sufficiently close to actuality for most practical purposes. Take the case we have just considered, namely, when X is a normal variate with mean θ and unit standard deviation. Values of y_0, y_h, Y_0, and Y_h were obtained for a number of values of h and θ using eqns (7.46) and (7.49) when $P(0)$ and $E(N; 0)$ were known by obtaining values for $g_N(Z_N; \theta)$ by quadrature or the method just described. Some of their values are shown in Table 7.4. These typify all of the values of θ in the range shown and for values of h between 3 and 5. It is clear that

- y_0 and Y_0 are almost equal to one another as are y_h and Y_h;
- y_0 (and therefore Y_0), $h - y_h$ (also $h - Y_h$) are nearly invariant with respect to the whole range of values of h considered.

Hence to obtain the ARLs of a variety of different cusum schemes the values of y_0, y_h, Y_0, and Y_h (together with many other calculations which have been carried out) indicate the possibility of taking y_0 to be constant and equal to 0.48. Furthermore, the values indicated in the table imply that $h - y_h$ and $h - Y_h$ are close to $0.22(3 + |\theta|)$. We can therefore ask what values of $L(0)$ do we obtain using these two approximations and simplified expressions based on eqns (7.42) and (7.45)?

The expressions we derive from (7.42) and (7.45) for $L(0)$ using the near equivalence of y_0 and Y_0 etc., do not depend particularly on the algebraic form of the cumulated variate. For this reason, and the implications of the central limit theorem it is reasonable to expect that the expressions which we now develop will apply to distributions of X other than the normal distribution

Table 7.4   Values of y_0, y_h, Y_0, and Y_h computed for X being $N(\theta; 1)$

h	θ	y_0	Y_0	$h - y_h$	$h - Y_h$
3	1.2	0.45	0.50	0.88	0.79
	0.40	0.51	0.53	0.74	0.72
	−0.60	0.52	0.47	0.77	0.86
	−1.20	0.45	0.41	0.88	0.99
5	1.20	0.45	0.50	0.96	0.85
	0.40	0.53	0.55	0.76	0.73
	−0.60	0.50	0.48	0.79	0.85
	−1.20	0.45	0.41	0.96	1.10

If we consider the dominant terms in eqns (7.42) and (7.45) and use the result

$$L(0) = E(N; 0)/[1 - P(0)],$$

we find that a close approximation to $L(0)$ should be given by,

$$\theta L_a(0) = \frac{\beta_h(y_h) + \{[\beta_0(0) - \beta_0(y_0)]F(0; \theta) + \beta_0(y_0)\}[\mu_h(y_h) - \beta_0(y_0)]}{1 - \mu_0(0)F(0; \theta) - \mu_0(y_0)[1 - F(0; \theta)]}$$

$$(7.67)$$

where,

$$\mu_0(y) = e^{-2\theta y}\Phi(\theta - y)/\Phi(-\theta - y)$$

$$\mu_h(y) = e^{-2\theta y}\Phi(y - h - \theta)/\Phi(y - h + \theta)$$

$$\beta_0(y) = y + \theta - \phi(-y - \theta)/\Phi(-\theta - y)$$

and

$$\beta_h(y) = y + \theta + \phi(y - h + \theta)/\Phi(y - h + \theta).$$

An example of the accuracy of approximation (7.67) and using $y_0 = Y_0 = 0.48$ with $h - y_h = h - Y_h = 0.22(3 + |\theta|)$ is illustrated in Table 7.5.

Equation (7.67) is particularly useful for obtaining values of $L(0)$ for θ between θ_a at AQL and θ_r at RQL, once a scheme has been designed to achieve specific values at these two quality levels. In view of these comparisons, use of eqns (7.67) and those for $\mu_0(0)$, $\mu_0(y_0)$, $\beta_0(0)$, and $\beta_0(y_0)$ leads to a very simple expression for $L_a(0)$, namely

$$\theta L_a(0) = y_h + b(\theta) + a(\theta)\exp(-2\theta y_h) \tag{7.68}$$

with

$$a(\theta) = \frac{\{[\beta_0(0) - \beta_0(y_0)]\Phi(-\theta) + \beta_0(y_0)\}\Phi[-\theta - 0.22(3 + |\theta|)]}{\{1 - \mu_0(0)\Phi(-\theta) - \mu_0(y_0)\Phi(\theta)\}\Phi[\theta - 0.22(3 + |\theta|)]}$$

Table 7.5 Values of $L_a(0)$ using $y_0 = Y_0 = 0.48$, $h - y_h = h - Y_h = 0.22(3 + |\theta|)$ compared with the correct values of $L(0)$ when $h = 3$

θ	$L(0)$	$L_a(0)$	θ	$L(0)$	$L_a(0)$
−0.40	74	75	0.40	7.43	7.61
−0.60	198	200	0.60	5.61	5.67
−0.80	585	588	0.80	4.49	4.50

and $b(\theta)$ is given by the expression

$$\theta + \{\phi[\theta - 0.22(3 + |\theta|)] - \beta_0(y_0)$$

$$\left/[1 - \mu_0(0)\Phi(-\theta) - \mu_0(y_0)\Phi(\theta)]\} \cdot \{\Phi[\theta - 0.22(3 + |\theta|)]\}^{-1}.\right.$$

Once values of $a(\theta)$ and $b(\theta)$ have been calculated for a range of values of θ, very close approximations to $L(0)$ can be easily obtained for a wide range of values of h using (7.68). If we take y_h to be $h - 0.22(3 + |\theta|)$ then

$$\theta L_a(0) = h + B(\theta) + A(\theta)\exp(-2\theta h) \tag{7.69}$$

with

$$A(\theta) = a(\theta)\exp[0.44(3 + |\theta|)] \tag{7.70}$$

$$B(\theta) = b(\theta) - 0.22(3 + |\theta|). \tag{7.71}$$

To illustrate the procedure together with the formulation of a cusum scheme with specified ARLs, at AQL, and RQL let us return to Example 6.3. Consider schemes where a process is sampled either twice or four times every twenty-four hours. When the control statistic is X with mean m and the values being cumulated are $(X - k)$ their mean is $\theta = m - k$. Let us denote their standard deviation by $\sigma(x)$ and use $L(0; \theta)$ for the ARL, of the control scheme. For $\theta = \theta_a$ at AQL, and $\theta = \theta_r$ at RQL, $L(0; \theta_a)$ for the first scheme has to be 240 for the run length to be 120 days and $L(0; \theta_a) = 480$ for the second. To compare the profiles of rules used in Chapter 6 with those of a cumulative sum scheme, take the sample size n to be 25. Since $m_a = 5.72$ and $m_r = 7.00$ take the reference value of the cusum scheme to be 6.36. Evidently $\theta_a = (m_a - k)/\sigma(x) = -0.80$, interpolation using the values of $\log L(0; \theta)$ in Table 7.28 gives $h/\sigma(x) = 2.88$ for the first scheme and $h/\sigma(x) = 2.45$ for the second. Values of $A(\theta)$ and $B(\theta)$ and $L(0; \theta)$ for both schemes are shown in Table 7.6. It shows that for an equivalent sample size the profile of the cusum scheme relating to sampling four times a day is better than that for the rule of Table 6.6. To achieve a rule with profile close to the latter n will be less than 25. It also indicates that a value less than 64 can be found for a cusum scheme with profile near to that for the scheme of Table 6.6 when two samples are taken daily. For such a scheme Tables 7.4 and 7.5 show that $n = 40$ with $h/\sigma(x) = 2.00$ and $k = 6.36$ has profile close to that of the double sampling scheme of Table 6.6.

Fitting polynomials in θ to computed values of $A(\theta)$ and $B(\theta)$ for values of θ most likely to arise in practice give the following closely approximating expressions for $L_a(0; \theta)$. For θ between -1.20 and -0.20

$$L_a(0; \theta) = (h/\theta) + [(3.35\theta^2 + 12.48\theta + 18)\theta + 12.09 + 4.89/\theta]$$

$$- [(0.78\theta^2 + 3.90\theta + 11.18)\theta + 8 + 4.83/\theta]\exp(-2\theta h), \quad (7.72)$$

Table 7.6   Values of $A(\theta)$, $B(\theta)$, and $L(0; \theta)$ for m between 5.72 and 7.13 when $k = 6.36$ $h/\sigma(x) = 2.88$ and 2.45 whilst $n = 25$

m	θ	$A(\theta)$	$B(\theta)$	$h/\sigma(x) = 2.45$	$h/\sigma(x) = 2.88$
5.72	−0.80	−3.9269	1.697	242.2	486.5
5.92	−0.55	−3.2376	1.917	79.2	131.1
6.12	−0.30	−3.3600	2.571	31.9	44.9
6.52	0.20	1.5664	−1.172	9.3	11.1
6.92	0.70	0.1469	0.432	4.1	4.7
7.00	0.80	0.0914	0.566	3.8	4.3
7.13	0.96	0.0533	0.673	3.2	3.7

Table 7.7   Values of $L_a(0; \theta)$ given by eqns (7.72) and (7.73)

	$h = 2$		$h = 2.5$		$h = 3.5$		$h = 4.0$
θ	$L_a(0; \theta)$	θ	$L_a(0; \theta)$	θ	$L_a(0; \theta)$	θ	$L_a(0; \theta)$
−1.20	625	−1.00	721	−0.70	675	−0.60	667
−1.10	404	−0.90	430	−0.60	361	−0.50	336
−1.10	263	−0.80	261	−0.50	200	−0.40	180
0.20	7.7	0.20	9.6	0.30	10.2	0.30	11.8
0.40	5.0	0.40	5.4	0.50	7.4	0.50	8.4
0.80	2.4	0.80	3.5	0.80	4.8	0.80	4.9

whilst for θ between 0.20 and 1.10

$$L_a(0; \theta) = (h/\theta) + [(5.02\theta - 13.01)\theta + 11.70 - 3.02/\theta]$$
$$- [(5.47\theta + 13.97)\theta + 11.82 - 3.40/\theta] \exp(-2\theta h) \tag{7.73}$$

and when $\theta = 0$,

$$L_a(0; \theta) = 2.4978 \exp[h(0.7930 - 0.0471h)]. \tag{7.74}$$

For values of $\theta > 1$ we can use the expression

$$L_a(0; \theta) = (h/\theta) + 0.70. \tag{7.75}$$

Some values of $L_a(0; \theta)$ given by these last equations are shown in Table 7.7. Comparisons with run lengths in Tables 7.29 and 7.30 illustrate that their accuracy is certainly acceptable for the design of practical schemes. These are accurate run lengths computed by using numerical methods of integration.

As already remarked, needless plotting of a cumulative sum chart can be achieved when sampled data is only used for control purposes. When we want to detect increases in the process mean from m_a we cumulate results only when a sample value greater than $k(> m_a)$ occurs. This and subsequent values of $(X_i - k)$ are cumulated until the sum returns below zero or exceeds h. We call this procedure a **decision interval** scheme. When we need to test for positive and negative deviations from m_a two reference values k_1 and k_2 are used and two decision interval schemes are run concurrently. Thus, if the process being controlled produces rejectable material when its mean level deviates from the value m_a by more than $\pm\Delta$, we could take $k_1 = m_a + \Delta/2$ and $k_2 = m_a - \Delta/2$. Action would then be taken if the cumulated sum using k_1 exceeds h or it becomes less than $-h$ using k_2.

An advantage of plotting decision interval schemes from a practical point of view, is that the charts which result do not run off the paper. The arithmetic of such schemes can, of course, easily be applied to a set of successive sample values without the need to plot a graph at all. However, as already pointed out, cusum charts can indicate points at which changes in the process may well have occurred. This feature of such charts is most useful from a research point of view when, as with many of the applications with which the author has been involved the causes of changes in level were unknown and needed to be discovered. A plot of recent results as a cusum chart can give an extemely useful indication of the time at which the change in level occurred.

7.18 Simplified design of cusum schemes

Once values of $L(0)$ are known for ranges of values of θ and h, we shall see that simplified procedures can be devised to design control schemes with specific practical requirements. To illustrate methods of doing so take the case when sampled values X_i are independent $N(m; \sigma^2)$ variates. The parameters which can be varied in the design of a scheme are the reference value k, the decision interval h and the number of items n tested per sample.

Suppose we want to control m and, for the reasons discussed in previous chapters, the control statistic we choose to use is the sample mean

$$\bar{X} = \sum_{i=1}^{i=n} X_i/n.$$

Denote the value of the ARL of a scheme when $m = m_a$ by $L(0; m_a)$ and that when $m = m_r$ by $L(0; m_r)$; furthermore, use $\sigma(x)$ for the standard deviation of the control statistic, in this case $\sigma/\sqrt{n}$. One way of determining values of h, k and n for specified values of $L(0; m_a)$ and $L(0; m_r)$ is to use a nomogram similar to the one shown in Fig 7.9. In this diagram $|\theta|$ of Tables 7.28 and 7.29 is $|k - m|/\sigma(x)$. The run lengths and values of θ given in Table 7.28 are used to draw the curve for $h/\sigma(x)$ and calibrate the vertical scale $L(0; m_a)$. The position of the curve for $L(0; m_r)$ is then obtained using the values that $|\theta|$ and $L(0; m_r)$ take for a number of specified values of $L(0; m_a)$. To obtain the values of $L(0; m_a)$ and $L(0; m_r)$ for specified h, k, and

Table 7.8 Values of $|\theta| = |k - m|/\sigma(x)$ for different $h/\sigma(x)$ and $L(0; \theta)$ for a normal variate

$L(0; \theta)$	$h/\sigma(x) = 2$	$h/\sigma(x) = 3$	$h/\sigma(x) = 4$	$h/\sigma(x) = 5$
1000	1.31	0.90	0.66	0.50
750	1.25	0.85	0.62	0.47
500	1.16	0.78	0.56	0.41
250	0.95	0.65	0.46	0.32
100	0.77	0.46	0.29	0.18

Table 7.9 Values of $|\theta| = |k - m|/\sigma(x)$ for different combinations of $L(0; \theta)$ and $L(0; \theta)$

$L(0; \theta_r)$	$L(0; \theta_a) = 250$	$L(0; \theta_r) = 500$	$L(0; \theta_a) = 1000$
2.5	1.05	1.17	1.27
3.0	0.94	1.035	1.13
4.0	0.78	0.85	0.92
5.0	0.68	0.74	0.80
6.0	0.60	0.655	0.71
7.5	0.52	0.57	0.62
10.0	0.43	0.48	0.52

$\sigma(x)$ we place a ruler on the nomogram joining the points $h/\sigma(x)$ and $|k - m|/\sigma(x)$. Some typical values which can be used to construct the nomogram of Fig 7.9 are given in Tables 7.8 and 7.9, θ_a and θ_r being values of θ at AQL and RQL.

Formal expressions we have considered to obtain values of $L(0; \theta)$ could lead to the conclusion that the design in practice of cusum schemes with specific run length profiles could be a somewhat complicated matter. Let us examine whether this is in fact the case. Clearly, there are number of schemes with different decision intervals and reference values with common run lenths at AQL, RQL, or both. An examination of these schemes indicates that there are advantages to be gained from using a central value for k. Thus for schemes designed to control a process mean, if m_a is its value at AQL, and m_r is that at RQL, we would take k to be $(m_a + m_r)/2$. We find that for a given value of n and ARL, at AQL, the ARL at RQL, is minimized for values of k in this region. Furthermore, a scheme with specified run lengths at both AQL and RQL has minimal n when $k = (m_a + m_r)/2$.

Graphs of sample size plotted against different k values show a shallow minimum in the region of central values of k. Provided the reference value is in this region there will be negligible differences in the sample sizes of comparable schemes. This feature is important from a practical point of view in the use of schemes on the factory floor or in clinical testing labs where it is sometimes advantageous to simplify the arithmetic necessary in its operation by rounding off the reference value. More

Table 7.10   Sample size comparison of a rule of a Shewhart and cusum scheme with similar run length profiles

m	Shewhart ARL $n = 20$	$(m - k)/\sigma(x)$	Cusum ARL $h = 2.03, k = 13.50$
12.00	180	−0.90	182.7
12.25	97	−0.75	99
12.50	55	−0.60	57
13.00	20	−0.30	22
14.00	4.4	0.30	3.9
15.00	1.8	0.90	3.0

importantly, however, the capability to specify k leads to a substantial simplification in the design of cusum schemes in practice. The design of a control scheme with specified ARLs, at AQL, and RQL, becomes a simple matter by using eqns (7.72) and (7.73) or nomograms like Fig. 7.9. Let us accordingly use these two expressions to compare the sample sizes required for a cusum scheme with similar profiles to the schemes discussed in Chapter 6. The Shewhart chart of Table 6.2 is designed for $L(0; m_a) = 180$ and $L(0; m_r) = 2.0$ when $m_a = 12.0$ and $m_r = 2.0$ with $\sigma = 5.0$ and $n = 20$. What reduction in n can we anticipate if the cusum method of control is used rather than a Shewhart chart if it is to have a similar profile? Table 7.10 shows that a scheme with $n = 9$ (over half the size) has profile almost equivalent to the scheme of Table 6.2.

Table 7.11 shows a similar reduction in sampling for nearly equivalent profiles of schemes of types 1 and 2. The table relates to Example 6.3 where we saw that when sampling 4 times a day, 100 samples were required each 24 hours to achieve the profile shown. The sequential method of control only needs a total of 48 samples. With eqn (7.72) and (7.73) we can examine the relationship between the sample size required by cumulative sum schemes with different sampling intervals. If for specified average run lengths at AQL, and RQL, we determine the values of h and n for schemes with different sampling intervals we find a nearly linear relationship between sample size and sampling frequency. An example is shown in Table 7.12 which relates to Table 6.5 and Example 6.3. Suppose we need the run length of the control schemes to be 120 days when $m_a = 5.72$ and 1.5 days when $m_r = 7.00$. The central reference value is 6.36. Since n has to be an integer it is convenient to adjust the ARL, at RQL, to 1.55. The table shows the values of run lengths given by using the nomogram of Fig. 7.9 when samples are taken every 4, 6, and 12 hours.

Values shown in Table 7.12 are typical of those found over a wide range of different cumulative sum schemes. We find very little difference in the daily samples required whether a sample is taken once, twice, or three times a day. Accordingly in order to detect changes in the operating level of a manufacturing process or a change in level of a clinical testing procedure it is best to sample as frequently as practical considerations permit.

Table 7.11 Sample size comparisons of rule type 2 and cusum schemes with similar profiles

		Cusum ARL	
m	$n = 25$	$n = 16, h = 3.53$	$n = 12, h = 4.02$
5.72	480	478	479
5.92	165	148	162
6.12	67.6	52.8	59.4
6.52	17.2	16.7	21.8
6.92	7.2	6.9	8.7
7.13	4.0	5.3	6.8

Table 7.12 Values of h and n for different sampling intervals

	ARL			
h	$|(k - m)|/\sigma(x)$	$m = 5.52$	$m = 7.00$	n
2.14	0.92	240	3.1	33
3.55	0.65	480	6.2	16
4.55	0.53	720	9.3	11

7.19 Run length distribution

The design of cusum schemes in practice is relatively easy with the use of equations like (7.72) and (7.73). We have seen that they give close approximations to the true ARL values. Their use can easily be justified in practice for a number of reasons. As already remarked it is important to appreciate that the formal distributions assumed for the cumulated statistic $S(X)$ are more often than not themselves approximations to reality. The comparisons we have discussed in Chapter 4 regarding the central limit theorem are examples in point. The distribution by eqn (4.27) for

$$S(X) = \left[\sum_{i=1}^{i=n} (X_i - \bar{X})^2 / (n - 3/2) \right]^{1/2}$$

is based on the assumption that individual values of X_i are normally distributed. There are many circumstances in practice where this is a reasonable assumption. It is none the less an approximation. The deviations of $L_a(0; \theta)$ from $L(0; \theta)$ based on equations like (7.72) and (7.73) fall well within the margin of error between $L(0; \theta)$ and the one based on the really true distribution of $S(X)$.

Further justification for the use of expressions with the accuracy of (7.72) and (7.73) follows when we consider the distribution of the run length $\mathbf{r}$ and its relationship to maintenance procedures usually employed in manufacturing industries or periodic routine checks carried out in the field of clinical testing.

Equation (7.35) shows that when $L(0; \theta)$ is large $\sigma(\mathbf{r}) \cong L(0; \theta)$. For this and other reasons it has been shown that at AQL the distribution of $\mathbf{r}$ is given by

$$p(\mathbf{r}; 0) \cong [L(0; \theta)]^{-1} \exp[-(\mathbf{r} - 1)/L(0; \theta)]. \tag{7.76}$$

Figure 7.7 shows a typical run length distribution at AQL. It gives this distribution of $\mathbf{r}$ when $\theta = -0.50$, $h = 3$, and X_i is $N(0; 1)$. The values of $p(\mathbf{r}; 0)$ were obtained using the values of $G_N(0; 0)$ of Table 7.1 and eqn (7.32). The figure also shows the negative exponential distribution of eqn (7.76) with the same ARL, namely 117. In the present context it is important to notice that the probability of obtaining runs less than the average is high. In this particular example the chance of obtaining a run as low as 20 or less is 0.1406; that is, odds of 1 in 7. The corresponding value given by (7.76) is 0.1496. It should be emphasized that this particular feature of the run length distribution is not confined to cumulative sum control schemes. It is, in fact, a common feature of the run length distributions of Shewhart type control schemes and control schemes in general and needs to be taken into account in the practical application of such schemes. In certain situations, for example when an erroneous out-of-control decision is expensive, it would be wise to take the out-of-control decision as a **warning** that a change in process or testing level may have occurred. Additional

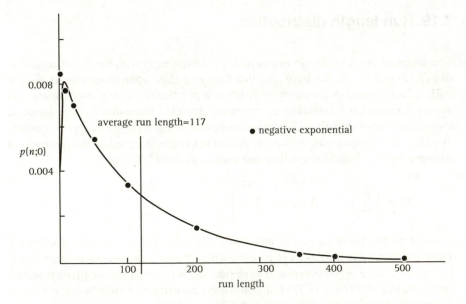

Fig. 7.7   Run length distribution for a scheme with $h = 3$ and sample drawn from a $N(1; 1)$ distribution.

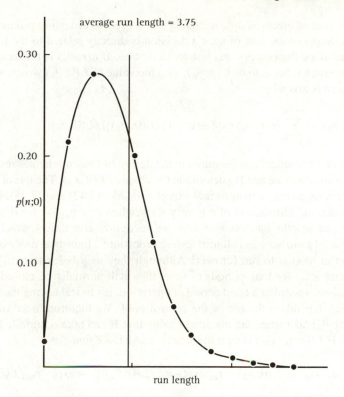

Fig. 7.8 Run length distribution for a scheme with $h = 3$ and samples drawn from a $N(1; 1)$ distribution.

testing could then be undertaken to confirm that a change in level has indeed taken place.

Equation (7.76) shows that when $L(0; \theta)$ is large the probability of getting a run length less than a specified multiple of $L(0; \theta)$ is nearly invariant since

$$\mathbf{r} = \lambda L(0; \theta)$$

$$\sum p(\mathbf{r}; 0) \cong 1 - \exp(-\lambda) \tag{7.77}$$

$$\mathbf{r} = 1.$$

Using this expression, the run length distribution is completely specified by $L(0; \theta)$ for values of θ in the region of AQL. Any other characteristic of a scheme for such quality values can accordingly be specified in terms of $L(0; \theta)$. We may, for example, wish to control the frequency with which run lengths less than a particular value arise. Values of $L(0; \theta)$ consistent with this constraint can be obtained from eqn (7.77). Suppose we require the frequency of run lengths less than 50 to be less than 1 in 10 at AQL, we find that $L(0; \theta_a)$ must be 476. We may, on the other hand, wish to control the cost of erroneously concluding that a process or testing procedure is out of control. Suppose the cost of such a decision is directly related to the time which has elapsed since the process was last set in motion. If $c(\mathbf{r}; \theta_a)$ is this cost function and if its average value is to be $C(\mathbf{r}; \theta_a)$, then the value of $L(0; \theta_a)$ which will ensure that this is so is given by

$$C(\mathbf{r}; \theta_a) = \sum_{r=1}^{r=\infty} \{c(\mathbf{r}; \theta_a) \exp[-(\mathbf{r} - 1)/L(0; \theta_a)]\}/L(0; \theta_a).$$

An important practical consideration in the design of cusum schemes relates to the accuracy with which we need to determine the value of $L(0; \theta_a)$. The use of equations like (7.72) is much easier than formal expressions like (7.42) and (7.45). Figure 7.7 illustrates that the distribution of $\mathbf{r}$ is very skew when $\theta = \theta_a$. We need to take this feature of run lengths into account once we recognize constraints which exist for most industrial processes and clinical testing procedures. Industrial processes are not usually set in motion to run for ever! Although they may be expected to do so at an in-control level for long periods of time, they will normally be closed down for routine maintenance after a fixed period of operation. In clinical testing the run length of tests will depend on the life of the control pool. We therefore need to see what values $L(0; \theta_a)$ take when the maximum value that $\mathbf{R}$ can take is curtailed. Suppose this is $\mathbf{r_m}$. If $L(0; \theta_a \cdot \mathbf{r_m})$ is then the schemes' ARL we find

$$L(0; \theta_a \cdot \mathbf{r_m}) \cong L(0; \theta_a) - \mathbf{r_m} \exp[-\mathbf{r_m}/L(0; \theta_a)]\{1 - \exp[-\mathbf{r_m}/L(0; \theta_a)]\}^{-1}.$$

$$(7.78)$$

Let us take a scheme for which $L(0; \theta_a) = 100$ and $\mathbf{r_m} = 250$. Using eqn (7.78) we find $L(0; \theta_a, 250) = 77.6$. We can obviously obtain a value for the run length' of the curtailed scheme which is closer to the one required by increasing the value of $L(0; \theta_a)$. If we take $L(0; \theta_a) = 200$ we then have $L(0; \theta_a, 250) = 99.6$. These latest calculations put into perspective the search for absolute accuracy in the determination of values of $L(0; \theta_a)$ on the one hand, and the influence on the actual value of the schemes' ARL, due to practical constraints on the other. It is clearly much more important to take into account reductions in the $L(0; \theta_a)$ value due to factors such as routine maintenance, or the limited life of a control pool, than the minor effects of the slight inaccuracies in the determination of $L(0; \theta_a)$ illustrated in Table 7.5.

A typical run length distribution at RQL of a scheme where the cumulated statistic is normally distributed with unit variance and $\theta_r = 1.00$ and $h = 3.00$ is shown in Fig 7.7. It illustrates that at this quality level the distribution of $\mathbf{R}$ has the desirable feature that the probability of a run less than $L(0; \theta_r)$ is high and the chance of getting one greater than $3L(0; \theta_r)$ is very small. For the scheme to which the figure relates $L(0; \theta_r) = 3.75$, the probability of a run less than 4 is 0.74 and the chance of getting a run greater than 11 is about 1 in 650. Again, from a practical point of view, we can be too fastidious with regard to the accuracy of $L(0; \theta_r)$. In this respect the calculation of $L(0; \theta_r)$ is based on the presumption that the particular sequential test which led to

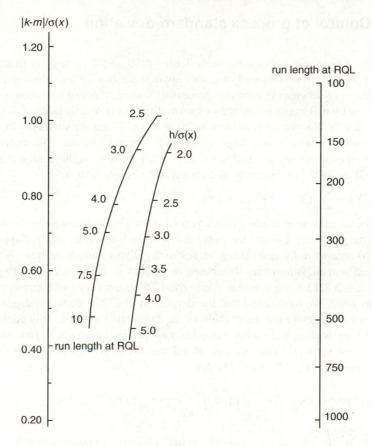

Fig. 7.9 Nomogram for ARL values when X_i is normally distributed.

the cumulated sum exceeding the decision interval commenced on the lower decision line. In practice this will not in general be the case. The point of change in quality level of the process being controlled will frequently occur whilst a cumulation is in progress. Thus the computed value of $L(0; \theta_r)$ on the above assumption is evidently greater than the one which will arise in practice.

The difference between $L(0; \theta_a)$ and $L(0; \theta_a, \mathbf{r})$ due to truncating the operation of a control scheme is similar for Shewhart charts and those described in Chapter 6, for the former we have seen that

$$p(\mathbf{r}) = q^{\mathbf{r}-1}p$$

and

$$L(0; \theta_a, \mathbf{r_m}) = L(0; \theta_a) - \mathbf{r_m}q^{\mathbf{r_m}}/(1 - q^{\mathbf{r_m}}). \qquad (7.79)$$

For a scheme with $L(0; \theta_a) = 100$ and $\mathbf{r_m} = 250$, eqn (7.79) gives $L(0; \theta_a \cdot \mathbf{r_m})$ equal to 78.

7.20 Control of process standard deviation

These latest observations together with Tables 4.10 and 4.12 suggest that from a practical point of view sufficiently accurate approximations for the values of $L(0; \theta_a)$ in the design of schemes to control a process or testing standard deviation σ should be obtained by making an assumption of normality for the distributions of $S_1(X)$ and in particular $S_2(X)$ when n is as large as 20 or more. There are situations in clinical testing, however, where such large values of n are not feasible. We shall shortly consider an example when n had to be very small indeed. We have seen that when individual values X_i are normally distributed but n is small, then

$$S_1(X) = \sum (X_i - \bar{X})^2 / (n - 1)$$

is distributed as a generalized gamma variate. Cox (1949) has also shown that the sample range closely follows the same distribution when n is small. Suppose we decide to control σ by cumulating values of $S_1(X)$ or sample ranges. When the cumulated statistic is normally distributed the values $y_0 = Y_0 = 0.48$ and $y_h = Y_h$ equal to $h - 0.22(3 + |\theta|)$ substituted into eqn (7.67) gave very close approximations $L_a(0)$ to $L(0)$. We have noted that the derivation of (7.67) does not depend upon any assumptions about the distribution of the cumulated statistic. It is accordingly reasonable to suppose that similar numerical approximations might be formulated for $y_0 = Y_0$ and $y_h = Y_h$ when the cumulated statistic is not normally distributed. It follows from eqns (7.22) and (7.11) that

$$P(y)E[\exp(\tau Z_N); y | Z_N \leq 0] = \int_{-\infty}^{0} \exp(\tau z_1) f(z_1 - y; \theta) dz_1$$

$$+ \sum_{n=2}^{n=\infty} \int_{z_n=-\infty}^{0} \int_{z_{n-1}=0}^{h} \mu_0(\tau; z_{n-1}) g_{n-1}(z_{n-1}; y) f(z_n - z_{n-1}; \theta) \, dz_{n-1} \, dz_n.$$

$$(7.80)$$

Similarly, from eqn (7.24) it follows that

$$P(y)E[Z_N; y | Z_N \leq 0] = \int_{-\infty}^{0} z_1 f(z_1 - y; \theta) \, dz_1$$

$$+ \sum_{n=2}^{n=\infty} \int_{z_n=-\infty}^{0} \int_{z_{n-1}=0}^{h} E(z_n; z_{n-1}) g_{n-1}(z_{n-1}; y) f(z_n - z_{n-1}; \theta) \, dz_{n-1} \, dz_n.$$

$$(7.81)$$

These equations suggest that y_0 should relate to the average value of z_{n-1} conditional on $z_n \leq 0$ and y_h to the average value of z_{n-1} conditional upon $z_n \geq h$. Rowlands and Nix (1981) observed that as n increases, the overshoot of a scheme above the upper decision line is evidently related to θ^+ where

$$\theta^+ = \int_{0}^{\infty} x f(x; \theta) \, dx / [1 - F(0; \theta)].$$

This observation led to the speculation that we might be able to find constants η and γ such that $y_h = h - \eta\theta^+$ and $y_0 = \gamma\theta^-$, where

$$\theta^- = \int_{-\infty}^{0} xf(x; \theta)\, dx / F(0; \theta)].$$

Calculations show that when the cumulated variate is normally distributed values of ARLs given by eqn (7.67) are very close indeed to their true values when $\eta = 1.5$ and $\gamma = 0.5$ at AQL, and these values are reversed at RQL. They have also been used for a wide variety of non-normal distributions. We find their use gives values of ARL, with 2 per cent error or less.

To illustrate the utility of these speculations, the close approximations to ARL they give, and their value in identifying optimal control schemes, let us consider the following example in which my colleagues and I were involved with the Tenovus Institute in Cardiff. It concerned the determination of specific hormone concentrations in plasma samples. Analysis of historical data revealed the necessity to control the standard deviation of test results as well as their mean. To achieve the former it was decided to insert replicated control samples into assays at regular intervals.

As we have seen, there are a number of methods of control we could adopt. We could use a single-sided Shewhart scheme, or one with warning and action lines, but, as has been pointed out, these make only limited use of the information in test data. They are particularly useful in the detection of sudden significant changes in the mean level of a test procedure or its standard deviation. Analysis of the available clinical data for the situation we were considering, indicated that sudden increases in the value of σ were unlikely. After maintaining a consistent in control value for some time the pattern of change was for σ to move out of control by steadily increasing in value. It is well known that cusum schemes are sensitive to the detection of small increases or decreases in the value of the parameter being controlled. Having regard to the very small value which in this case n could take and the need to use information in the test data as efficiently as possible, the use of a cusum method of control was certainly justifiable. Its use in clinical testing provides an additional and important feature. When a Shewhart chart is used it indicates action at the Nth sample after testing commenced. This could result in just checking the testing system only for the time the Nth batch was assayed. In doing so attention could become centred on this particular instant in time. In the author's not inconsiderable experience in the clinical testing field this can and does happen. This is particularly so when, as is often the case, laboratory staff have little knowledge of the principles on which statistical method is based. Clearly testing could have been off or moving off target much earlier than when the Nth sample was taken. Complete reliance on Shewhart type charts can result in an unknown number of incorrectly assayed patient tests. The cusum procedure suffers to a far less extent from this disadvantage. When differences between sample values and the specified acceptable quality level are cumulated, sequences of positive values of this sum begin to occur more frequently as the mean of these values exceeds AQL. These cumulated sums obviously contain information which can be used to indicate the point of change as well as the change in level which has occurred. Merely plotting the cumulated differences between sample values and the AQL value provides a useful visual aid to detecting the time when changes in level began to occur.

The present example also emphasizes remarks made in Chapter 1 regarding the association between appropriate decision criteria and fields of application. For a number of industrial processes best schemes will be those which for a fixed ARL at AQL minimize ARLs as soon as the process moves off target. In the clinical field there are situations, such as the present one, where replication is severely restricted and it is necessary to insist upon a small value of ARL at RQL, and at the same time maximize the life of the control pool when testing is on target. In these circumstances if two schemes A and B have a common ARL at RQL, A will be preferred to B if it has a larger ARL, when testing is on target. Let us now examine this criterion in the context of clinical quality control. For many industrial processes characteristics required of control schemes can be achieved by determining appropriate sample sizes. The cost of increasing sample size is frequently a secondary consideration. In clinical assays this is often not the case. Control material such as a plasma pool can be costly to prepare. In addition testing samples from it can be expensive, complicated and time consuming. In these circumstances there is a need to make control material last as long as possible. We are accordingly led to examine ways of improving the characteristics of control schemes by devices other than sample size adjustment. Duplicate testing is often the maximum replication that can be envisaged. If we increase replication r from 2 to 3 keeping sampling frequency constant, we reduce the life of the control pool by one-third. Circumstances exist where the consequences of such a reduction is both expensive and for technical reasons undesirable. The need to make the most effective use of small amounts of control material is clear. We can do so by selecting

(i) a testing technique which is sensitive to changes in the level of the feature being controlled;

(ii) parameter values which optimize values of criteria on which the choice of a technique is based;

(iii) a function of the test data which additionally gives optimal values for the design criteria of the selected technique.

The selection of the optimal scheme is based on fixing the ARL at RQL and maximizing it at AQL. Identifying such schemes is clearly a constrained maximization problem which is difficult to solve without simple and sufficiently accurate formulae with which to evaluate average run lengths. Without such expressions it would be necessary to resort to a numerical search procedure to identify an optimal scheme. This would entail a separate ARL calculation for each cusum scheme considered. Numerical evaluation of ARLs is time-consuming and not suitable for the routine determination of best schemes. We have seen that the standard deviation of the run length distribution approximates at AQL to its average value $L(0)$. Thus, if $L(0)$ is to be obtained by simulation over 156 000 scores need to be generated to achieve a standard error of 10 for values of $L(0)$ as low as 250. To achieve informed simulation or quadrature at a feasible computational level, the need to devise formal expressions which give tolerably accurate ARLs is clear. Such expressions can be used instead of nomograms which usually only cover a limited range of values for $L(0)$ and which can give values which are not particularly accurate. In many practical situations it would be a considerable advantage to be able to compute features of cusum charts using arithmetic expressions approaching the simplicity of those required for Shewhart schemes. Exploratory computations indicate that eqn (7.67) can be used to derive

(a) easily programmed expressions which give very close estimates of $L(0)$ for a wide range of values of h and θ;

(b) simple approximating expressions which do not require sophisticated computing equipment to obtain close estimates of $L(0)$.

The need to distinguish between (a) and (b) arises from the different demands of theory and practice. Expressions with the accuracy of (a) are necessary to confirm the existence of optimal schemes. Those of (b) relate to situations where the availability of reasonably good approximating expressions is invaluable from the point of view of designing and operating schemes in practice. We achieve (a) by computing values for y_0 and y_h with the values of η and γ indicated and (b) by putting $y_0 = 0$ in eqn (7.67), which gives

$$L_a(0) = \beta_h(y_h) + \beta_0(0)[\mu_h(y_h) - 1]/[1 - \mu_0(0)]. \tag{7.82}$$

Returning to the particular example under consideration, data indicated that the sampled values (functions of hormone levels) could be assumed normally distributed. Let us denote the values obtained for the replicates of the ith control batch by $X_{1i} \cdot X_{2i}, \ldots, X_{ri}$, then X_{ji} is an independent $N(m; \sigma^2)$ variate and the parameter being controlled is σ. The value of r is to be 2 or 3. It was not obvious to us when we approached this problem which control statistic would lead to the most efficient monitoring scheme, or what degree of change in the value of $L(0)$ we could anticipate by increasing r from 2 to 3, or indeed how accurate values of $L(0)$ would be taking $\eta = 1.5$ and $\gamma = 0.5$. It was decided to consider statistics proportional to $S(X)$ and $S^2(X)$ where

$$S^2(X) = \sum_{j=1}^{j=r} (X_{ij} - \bar{X}_i)^2/r \quad \text{with } \bar{X}_i = \sum_{j=1}^{j=r} X_{ij}/r,$$

so that $S^2(X)/\sigma^2$ is distributed as χ^2 with $(r - 1)$ degrees of freedom. Let us use σ_T to denote the value of σ when testing is on target and $\alpha\sigma_T$ when it is not. From eqn (4.16), if we write $S^2(X)/\sigma_T^2 = Z$ and $S(X)/\sigma_T = Y$, then

$$f(z) = \{2^{(r-1)/2}\Gamma[(r-1)/2]\}^{-1}(r/\alpha^2)^{(r-1)/2}z^{(r-3)/2}\exp(-rz/2\alpha^2)$$

$$\tag{7.83}$$

and

$$f(y) = \{2^{(r-3)/2}\Gamma[(r-1)/2]\}^{-1}(r/\alpha^2)^{(r-1)/2}y^{(r-2)}\exp(-ry^2/2\alpha^2), \quad (7.84)$$

so that $E(Z) = \alpha^2/2$ and $E(Y) = \alpha/\sqrt{\pi}$ when $r = 2$ and $E(Z) = 2\alpha^2/3$ with $E(Y) = \alpha(\pi/6)^{1/2}$ when $r = 3$.

The truncated distributions of these two statistics with the singularity of the frequency function of $S^2(X) - k$ illustrate the need to formulate expressions which can be used to compute $L(0)$ which do not depend on quadrature. Examples of the accuracy obtained using $L_a(0)$ given by (7.67) and the specified values of η and γ are shown in Tables 7.13 and 7.14, where $r = 2$. Values for $L(0)$ also given in these tables were obtained by quadrature when possible and simulation when not. For variate Z when $r = 2$ we have

$$\mu_0(y) = \frac{[\Phi\{[2(k-y)]^{1/2}\exp(-t_0k)/\alpha\} - 1/2]\exp(t_0y)}{\Phi\{[2(k-y)]^{1/2}/\alpha\} - 1/2} \quad (y \le k) \quad (7.85)$$

and for $\mu_h(y)$ we obtain

$$\mu_h(y) = \frac{[1 - \Phi\{[2(h+k-y)]^{1/2}\exp(-t_0k)/\alpha\}]\exp(t_0y)}{\Phi\{-[2(h+k-y)]^{1/2}/\alpha\}} \quad (7.86)$$

whilst

$$\beta_0(y) = \alpha^2/2 + y - k - \alpha[(k-y)/2]^{1/2}\phi\{[2(k-y)]^{1/2}/\alpha]\}$$
$$/[\Phi\{[2(k-y)]^{1/2}/\alpha\} - 1/2] \quad (7.87)$$

and

$$\beta_h(y) = \alpha^2/2 + y - k - \alpha[(h+k-y)/2]^{1/2}\phi\{[2(h+k-y)]^{1/2}/\alpha]\}$$
$$/[\Phi\{-[2(h+k-y)]^{1/2}/\alpha\}] \quad (7.88)$$

where t_0 is the non-zero root of $\exp(-\tau k) = (1 - \alpha^2\tau)^{1/2}$. Table 7.13 shows the values of $L_a(0)$ and those for $L(0)$ which were obtained with simulation. The value of r was 2. The simulation estimates are based on 10^5 runs and the figures in parentheses are the standard errors of them.

A similar calculation was carried out for schemes which cumulated $(Y - k)$. For these we find that when $r = 2$

$$\mu_0(y) = \frac{[\Phi\{[(k-y-t_0\alpha^2/2]\alpha^{-1}\sqrt{2}\} - \Phi(-t_0\alpha/\sqrt{2})]\exp(t_0y)}{2\Phi(t_0\alpha/\sqrt{2})\{\Phi[(k-y)\alpha^{-1}\sqrt{2}] - 1/2\}} \quad (7.89)$$

Table 7.13   Values of $L_a(0)$ and $L(0)$ obtained by simulation for schemes which cumulate $(Z - k)$ and $r = 2$

h	k	Simulation	$L_a(0)$	α
4.375	0.875	292 (0.91)	291	1.0
5.00	0.8125	371 (1.17)	367	1.0
4.375	0.875	5.96 (0.02)	5.93	2.0
5.00	0.8125	6.28 (0.02)	6.25	2.0

Table 7.14   Values of $L_a(0)$ and $L(0)$ when the cumulated statistic is $(Y - k)$ and $r = 2$

h	k	Quad $L(0)$	Sim $L(0)$	$L_a(0)$	α
1.0	1.0	119	123 (2.4)	117	1.0
1.5	0.75	81	79 (1.6)	81	1.0
1.0	1.0	4.96	4.90 (0.04)	5.03	2.0
1.5	0.75	9.58	9.63 (0.12)	9.98	1.5

whilst

$$\mu_h(y) = \frac{[\Phi[t_0\alpha/\sqrt{2} - (h + k - y)\alpha^{-1}\sqrt{2}]]\exp(t_0 y)}{2\Phi(t_0\alpha/\sqrt{2})\Phi[(y - k - h)\alpha^{-1}\sqrt{2}]} \tag{7.90}$$

$$\beta_0(y) = \frac{y - k + \alpha\{[\sqrt{(2\pi)}]^{-1} - \phi[(k - y)\alpha^{-1}\sqrt{2}]\}}{(\sqrt{2})\{\Phi[(k - y)\alpha^{-1}\sqrt{2}] - 1/2\}} \tag{7.91}$$

and finally

$$\beta_h(y) = \frac{y - k + \alpha\phi[(h + k - y)\alpha^{-1}\sqrt{2}]}{(\sqrt{2})\Phi[(y - k - h)\alpha^{-1}\sqrt{2}]} \tag{7.92}$$

where t_0 is the non-zero root of

$$\exp(k\tau - \alpha^2\tau^2/4) = 2\Phi(\tau\alpha/\sqrt{2}).$$

Values of $L_a(0)$ using eqns (7.89) to (7.92) with (7.67) are given in Table 7.14. In this instance, it is possible to solve the standard integral equations for run lengths numerically, so that calculations of $L(0)$ thereby obtained are also given. In addition, simulated values were obtained to validate the programs and random number generators being used.

The values given in Tables 7.13 and 7.14, together with many other calculations, confirm that $L_a(0)$ obtained from eqn (7.67) can be used to establish the existence of optimal control schemes. An ARL subroutine was therefore written based upon values

Table 7.15 Values of $L_a(0)$ and $L(0)$ when the cumulated statistic is $(Y - k)$ and $r = 3$

h	k	Quad $L(0)$	Sim $L(0)$	$L_a(0)$	α
1.25	0.975	192	190	193	1.0
1.50	0.960	331	334	334	1.0
1.25	0.975	8.04	8.00	8.35	1.5
1.50	0.960	9.33	9.38	9.68	1.5

Table 7.16 Parameters of optimal schemes for duplicate samples with control statistics S^2/σ_T^2 and S/σ_T

ARL at $\alpha\sigma_T$	Control statistic S^2/σ_T^2			ARL at σ_T	Control statistic S/σ_T		ARL at σ_T
	α	Opt k	Opt h		Opt k	Opt h	
5	2	0.907	3.219	130	0.993	1.004	114
6	2	0.902	4.302	308	0.980	1.266	241
7	2	0.915	5.394	714	0.970	1.510	485
8	1.75	0.824	4.340	241	0.954	1.342	181

of $L_a(0)$ calculated in this way, and a maximum search procedure was run using it. The method of simulated trials was also used to check on a number of optimal or near optimal schemes.

Optimal control schemes which relate to the example we are considering, are shown in Tables 7.13 to 7.16 for $r = 2$ and $r = 3$. From them it is clear that S^2/σ_T^2 has higher run lengths at AQL for a given run length at RQL and that the run lengths which can be achieved when $r = 3$ is much greater than when $r = 2$.

With regard to our example, the following condition was laid down, namely, if the standard deviation of testing deteriorates to $2\sigma_T$ then we need, on average, to detect this change in two days. This was to ensure that patients needing to be reassessed would in general only be those tested over a two day period. Two strategies were available with regard to protecting the life of the control pool. For example we could insert duplicate tests into routine testing 3 times a day or we could insert triplicate tests into the procedure twice a day.

We can use the values given in these tables to decide whether there is any advantage to be gained by the choice of S^2/σ_T^2 or S/σ_T as the control statistic. For $r = 2$ the ARL when $\alpha = 2$ needs to be 6. For this run length Table 7.16 gives the maximum achievable average run length to be of the order of 308 when S^2/σ_T^2 is used and 241 when S/σ_T is used. These two schemes give anticipated runs in days of 103 and 80 when testing is on target. For triplicate sampling the same life of the control pool requires an ARL of 4 when $\alpha = 2$, Table 7.17 gives the maximum ARL to be very

Table 7.17 Parameters of optimal schemes for triplicate samples with control statistics S^2/σ_T^2 and S/σ_T

ARL at $\alpha\sigma_T$	Control statistic S^2/σ_T^2				Control statistic S/σ_T		
	α	Opt k	Opt h	ARL at σ_T	Opt k	Opt h	ARL at σ_T
4	2.00	1.229	3.599	559	1.132	1.101	556
5	1.75	1.103	3.282	260	1.064	1.062	244
6	1.75	1.105	4.218	689	1.061	1.301	584
7	1.75	1.106	5.151	1794	1.044	1.582	1338

nearly the same for both statistics, namely, 550 when testing is on target. The run length in days would therefore be 275; that is, over twice that which can be achieved using the best scheme for the duplicate test procedure.

We see that values given in Tables 7.18 and 7.19 validate the simulation procedures used in this example and the comparisons we shall shortly consider with regard to the properties of combined cusum control schemes. Before considering these it is important, however, to draw attention to an overriding practical consideration which has received little attention in recent research. We find, and both tables illustrate, that for nearly optimal schemes we do not need accurate determinations of h and k. These tables and many other calculations indicate that values of h and k over a wide range have nearly equivalent on target average run lengths. Equation (7.93) gives an expression due to Regula which can be used to determine the optimal value of k for a cusum scheme. Much attention has been focused on the identification of optimal schemes. Its significance must be seriously considered from a practical point of view as well as a theoretical one. To illustrate this observation take the schemes in the last rows of Table 7.19 which shows that for values of h equal to 1.434 and 1.73 with k equal to 0.898 and 0.959, the difference in $L_a(0)$ is 16. This difference and others in the table may be important from a theoretical point of view with regard to establishing the existence of optimal schemes and legitimate comparisons between the use of different control statistics. However, from a practical point of view the assumptions on which the calculations of ARLs are based, and the effects of routine industrial maintenance, or the finite life of control pools in clinical testing, do not justify searches for precise values of h and k to design practical schemes with exactly determined run lengths.

If the variate being cumulated is X, it has been shown that for some distributions of X the optimal reference value is given by the solution of the equation

$$f(x; m_a)/f(x; m_r) = 1, \tag{7.93}$$

where m_a and m_r are the parameter values for X which correspond to AQL and RQL. The value of k given by writing $k = x$ in this equation is $(m_a + m_r)/2$ when X is normally distributed. This accords with the value of k used in our earlier considerations of cusum schemes for the control of the mean of a normally distributed variate.

If we write $k(r; z)$ and $k(r; y)$ for the values of k for the variates Z and Y with the distributions of eqns (7.83) and (7.84), we find that

$$k(r; z) = [2(r - 1)\alpha^2 \ln \alpha]/[(\alpha^2 - 1)/r] \tag{7.94}$$

and

$$k(r; y) = [k(r; z)]^{1/2}. \tag{7.95}$$

The values of k given by these two equations are shown in Tables 7.18 and 7.19.

The steady growth in the application of statistical methods of quality control in clinical testing has been accompanied by the publication in non-statistical journals of comparisons of various quality control techniques. These have frequently been carried out using questionable simulation procedures, probabilistic criteria which are not particularly relevant to clinical testing, and comparisons which do not compare the best which can be achieved with one technique with the best which can be achieved with another. In addition to comparisons between individual schemes, consideration has also been given to the combination of different methods of control. These have been proposed in attempts to design procedures which utilize the best features of each constituent test, the theory being that an appropriate combination may have superior properties to those of its separate parts. It can be argued, however, that some combinations which have been suggested would be complicated to operate and interpret in practice. Furthermore, control schemes which have been regarded by the authors of these publications as different from one another, are in fact not as different as they appear to believe. Thus the discussions of the relative merits of cusum and Shewhart type charts leads to the impression that their authors look on these two as quite different techniques. We have seen, however, that some Shewhart charts with warning and action lines are cusum schemes with particular if somewhat crude scores assigned to sample values.

We need to identify a base line which can be used to compare individual techniques or combinations of them. To do so, let us return to our first concept of best schemes. Namely, for two tests A and B, with common ARL at AQL, A is considered conditionally better over a range of values of θ if for this range the ARLs of A are less than B. We use this definition to define the concept of envelopes. These provide a suitable foundation on which to judge the potential of different scheme combinations. Westgard *et al.* (1981) have examined schemes which combine cusum and Shewhart charts, whilst Bissell (1969) has investigated the superimposition of some five different cusum schemes and Lucas (1973) has considered a mask with a parabolic nose. Neither of these studies were carried out using the envelope concept. When this is done, we find that control schemes which are close to the best that can be achieved (certainly from a practical point of view) consist of an appropriate combination of just two cusum control rules.

Let us use $S(a, b)$ to denote a Shewhart scheme with warning line (or lines) distance a, and action line (or lines) distance b from the mean in-control value. Using the notation of Van Dobben de Brun (1968) denote the cusum scheme with decision interval h and reference value k by $C(h, k)$. $S(b)$ signifies a Shewhart scheme with action lines only distance b from the line representing the in-control value (that is, $\theta = 0$).

If two control schemes with a common on target ARL are such that the run length profile of the first lies entirely below that of the second for $\theta \neq 0$, then the first scheme will be said to be uniformly better than the second. If for a given on target

Table 7.18   Values of $h, k, L_a(0)$ and simulated $L(0)$ for a number of optimal or near optimal schemes for S^2/σ^2

r	α	$k(r; z)$	k	h	Off target ARL	$L_a(0)$	$L(0)$
			0.8636	3.00	10	99.2	98.3
2	1.5	0.7298	0.7258	3.56	"	102.9	102
			0.6426	4.00	"	100.0	99.7
			1.0705	3.75	6	289.5	290.2
2	2.0	0.9242	0.9197	4.25	"	302.9	303.0
			0.7926	4.75	"	287.9	288.3
			1.0098	2.90	7	129.2	127.9
3	1.5	0.9731	0.9692	3.059	"	130.0	128.1
			0.9126	3.30	"	128.1	126.6
3	1.5	0.9731	1.0117	3.90	9	335.4	331.0
			0.9707	4.13	"	339.0	335.0
			0.9418	4.30	"	336.9	332.0

Table 7.19   Values of $h, k, L_a(0)$ and simulated $L(0)$ for a number of optimal or near optimal schemes for S/σ

r	α	$k(r; y)$	k	h	Off target ARL	$L_a(0)$	$L(0)$
			0.8864	1.125	10	87.1	88.5
2	1.5	0.8543	0.8274	1.277	"	87.6	88.4
			0.7942	1.375	"	87.1	87.6
			1.0636	1.125	6	259.4	261
2	2.0	0.9614	0.9844	1.285	"	264.8	263
			0.9451	1.375	"	263.3	261
			1.0605	0.866	7	113.4	115
3	1.5	0.9864	0.9784	1.074	"	117.3	118
			0.9518	1.155	"	116.5	117
			1.0352	1.155	9	264.1	260
3	1.5	0.9864	0.9591	1.434	"	277.3	274
			0.8981	1.732	"	261.2	260

ARL we can identify a scheme with run length profile below all other alternatives it will be said to be uniformly best.

Minimal run length envelope

In terms of ARL, a scheme is optimal for a particular value of $\theta (\neq 0)$ if its $L(\theta)$ value is smaller than that of any other scheme having the same $L(0)$. If we denote this minimum value by $L_{\min}(\theta)$, then for a given $L(0)$ the minimal run length envelope is the curve of $L_{\min}(\theta)$ plotted against θ. Clearly, no scheme exists whose run length profile falls below this envelope at any point. Minimal run length envelopes provide an appropriate yardstick to assess the extent to which specific schemes are superior to all alternatives to them. Although we may not be able to identify minimal run length envelopes we can compute subminimal ones which on intuitive grounds must be close if not coincident with them. We can then identify schemes which can be regarded as best for all practical purposes.

Some comparisons (Westgard) have been based on test power. In these circumstances where probability may be more appropriate than run length, we can define power envelopes to assess different schemes. To do so, use $P(\theta)$ for the probability of reaching an out-of-control decision and $P_{\max}(\theta)$ for the maximum value that $P(\theta)$ can take for a specified $P(0)$. Then for a specified value of this probability a suitable criterion for best schemes would be the maximum power envelope, namely $P_{\max}(\theta)$ plotted against θ.

Cusum run length envelope

We can use $C(h^*; k^*)$ to represent the cusum decision interval scheme which for a given $L(0)$ has minimal $L(\theta)$, namely $L^*(\theta)$, for detecting a shift size θ away from target. The curve of $L^*(\theta)$ plotted against θ is the cusum run length envelope. Does this envelope coincide with the minimal envelope? Some results of Roberts (1966) raise the possibility that the ARL profile of a procedure based on a test of Girshick and Ruben violates the cusum envelope. However, Roberts obtained ARLs of the Girshick–Ruben's scheme under the assumption that a shift in the level of the parameter being controlled occurs some time after inspection starts. Consequently his results are not strictly comparable with cusum run length values based on the assumption that the shift occurred before inspection commenced. If we lift this restriction, a result of Lorden (1971) implies that the cusum envelope and the minimal envelope coincide when $L(0)$ is large. The question remains open but we can speculate that the cusum envelope is subminimal and accordingly provides a useful yardstick for the comparison of ARL profiles.

In clinical testing the statistical distributions of assay measurements are frequently found to be approximately normal or log normally distributed. Any dependence between test results can largely be removed by adjusting the sampling interval to ensure adequate separation of consecutive control samples. In the comparisons which now follow, we shall assume that the variate values plotted on $S(a; b)$ and $C(h; k)$ charts are statistically independent and are normally distributed with mean θ and unit standard deviation.

Envelope identification

Under these assumptions it is relatively easy to identify cusum run length envelopes. We have seen from eqn (7.93) that $k^* = \theta/2$. The envelope is therefore the curve of $L(\theta)$ for $C(h^*; \theta/2)$ plotted against θ. For given $L(0)$ the value of $L^*(\theta)$ can be obtained either from

(1) nomograms, for example those described by Goel and Wu;

(2) quadrature, to solve the integral equations for $P(0)$, $N(0)$ or $L(0)$;

(3) use of formal expressions such as eqn (7.67).

When used by itself the standard $S(b)$ chart is insensitive for the detection of changes in mean levels which are small relative to the standard deviation σ of sampled values. It does not utilize patterns of these which could be indicating a change in process or test level. The use of $S(a; b)$ schemes is an attempt to overcome these shortcomings. The superiority of cusum schemes for small changes in mean level and that supposed for Shewhart schemes for those with changes in level greater than 2σ, led to the search for schemes which combine the best features of both methods. Following Westgard, we could envisage the simultaneous use of both schemes. A simple Shewhart scheme widely used in clinical chemistry has action lines located at $\pm 2.5\sigma$. Its value at $\theta = 0$ is 81. Table 7.21 gives the ARL profile of this scheme and the corresponding cusum envelope for this on target run length. From the table we see that it is not the case that we cannot devise cumulative sum procedures with superior properties to Shewhart ones for changes in mean level in excess of 2σ. It is only that we cannot design a single cusum scheme with superior properties for changes in mean level less than 2σ and greater than it. Accordingly an alternative to the mixture of cusum and Shewhart schemes is the combined use of two appropriate cusum schemes. The envelope yardstick enables us to assess their use in practice.

The equivalence of some $S(a; b)$ schemes to cusum schemes with scores assigned to sample values enables us to anticipate superior run length features of combinations of cusum schemes. This, and similarities between the best profiles of $S(a; b)$ schemes,

Table 7.20 Comparison between the ARL profile of the Shewhart scheme with $L(0) = 81$ and the corresponding cusum envelope

	Deviation from target θ							
Scheme	0	0.5	1.0	1.5	2.0	2.5	3.0	4.0
$S(2.5)$	81	41.5	14.9	6.3	3.2	2.0	1.45	1.11
Cusum	81	18.1	7.0	3.9	2.5	1.8	1.40	1.07

leads to the implication that simple cusum scheme combinations exist which are easy to operate and interpret, and which in practical terms differ little from the best that can be achieved. To verify that this is so we need to obtain $C(h; k)$, $S(a; b)$ combinations which are best. Accordingly we need to select $S(a; b)$ schemes with minimum run lengths for $\theta \geq 2\sigma$. We therefore consider those schemes which utilise single sample values outside action lines and make sensible use of runs of results (2 or 3) between warning and action lines. We have seen that $S(a; b)$ schemes which fall into this category are equivalent to cusum schemes in which x_i values are replaced by crude scores. One of the schemes in such a $C(h; k)$, $S(a; b)$ combination does not therefore make the most efficient use of the information in sampled data. An obvious alternative is the combination of two $C(h; k)$ schemes. The first which minimizes $L(\theta)$ values when $0 < \theta \leq 2\sigma$ and the second which does the same for $\theta > 2\sigma$. In view of the comparisons which have been made in the clinical literature it is perhaps worth repeating the somewhat obvious remark that valid conclusions can only be drawn by comparing the best which can be achieved with $C(h; k)$, $S(a; b)$ combinations with the best that can be obtained by combining two cusum schemes. In view of the above observations we would not expect the run length or power characteristics of the best $C(h; k)$, $S(a; b)$ scheme to be superior to a properly combined pair of $C(h; k)$ schemes. In practice $C(h; k)$, $S(b)$ combinations might be reasonably simple to operate, Table 7.20, however, indicates that such a combination would only be justified for situations where there is ambivalence with regard to departures from target less than 2σ.

$S(a; b)$ envelope

If we select one of the rules discussed in Chapter 6 and compute its envelope, calculations indicate that this although subminimal should be close to the $S(a; b)$ envelope. A number of computational studies which have been undertaken support this conclusion. For the comparisons which follow let us accordingly compute the envelope for a rule of type 3. This scheme is equivalent to a cusum scheme with a score of -1 assigned to sample values between warning lines, $(N - 1)$ for those between warning and action lines and N for sample values outside an action line. For deviation θ from target denote the probability of score -1 by $p_0(\theta)$ and that for $(N - 1)$ by $p_1(\theta)$, then when $N = 4$

$$L(\theta) = \{1 + p_1(\theta) - p_0(\theta)[1 + p_1(\theta)^3]\}/[1 - p_0(\theta)]\{1 - p_1(\theta)[1 + p_1(\theta)]\}.$$
(7.96)

With this equation we can determine values of $p_0(\theta)$ and $p_1(\theta)$ which for a specific $L(0)$ yield the minimum $L(\theta)$ for given θ. Values for $L(0)$ equal to 100, 200 and 300 with X_i normally distributed and $\sigma = 1$ are shown in Table 7.21, as are values of the cusum envelope using the Goel and Wu nomogram.

The table indicates that when we need rapid detection of movements off target we should seek combinations of cusum schemes rather than cusum Shewhart ones.

Table 7.21 $L(\theta)$ values for $S(a; b)$ and $C(h; k)$ envelopes for a normal variate

Envelope		Deviation θ from target					
	0.00	0.50	1.00	1.50	2.00	3.00	4.00
S(a;b)	100	46.4	14.5	5.9	3.2	1.48	1.08
C(h;k)	100	19.3	7.4	4.0	2.6	1.44	1.08
S(a'b)	200	81.8	21.7	7.8	3.9	1.67	1.12
C(h;k)	200	24.2	8.7	4.6	3.0	1.59	1.12
S(a;b)	300	115.7	27.1	9.2	4.4	1.80	1.16
C(h;k)	300	27.2	9.5	5.0	3.2	1.68	1.16

7.21 Combined decision interval schemes

We have seen that the use of a decision interval scheme is equivalent to the use of a V-mask on a cusum chart with lead distance d, and semi-angle φ, with h equal to $d \tan \varphi$ and $k = \tan \varphi$ when $\omega = 1$. As already remarked, Barnard (1959), Lucas (1973), and Bissel (1979) suggested changing the shape of the mask in order to obtain schemes which deal effectively with both small and large shifts from target. A parabolic mask is one such suggestion. This and indeed any other curvilinear mask is equivalent to a polygon mask for cusum charts where sample values are plotted at equidistant points in time. Furthermore, a polygon mask is equivalent to the superimposition of a number of V-masks. The question which therefore arises is, do polygon masks lead to schemes which from a practical point of view are better than just one V-mask? If so how many component masks do we need and could they be designed easily?

Let us use $C_n(\mathbf{h}, \mathbf{k})$ to denote the constituent decision interval schemes of a polygon mask with decision intervals $\mathbf{h} = (h_1, h_2, \ldots, h_n)$ and with reference values $\mathbf{k} = (k_1, k_2, \ldots, k_n)$. The $h_1, h_2, \ldots, h_n$ are ordered so (h_1, k_1) refers to the parameters of the V-mask whose semi-angle is the most obtuse, namely, the one at the nose of the polygon.

Snub-nosed masks

In this notation $C_2(\mathbf{h}, \mathbf{k}) \equiv C(h_1, h_2; k_1, k_2)$ is a snub-nosed V-mask which is illustrated in Fig 7.7, here $\omega = 1$, $h_2 > h_1$, and $k_1 > k_2$. Such schemes are a combination of two separate cusum schemes. We expect the ARL profiles of these combinations to lie above the cusum envelope. Let us use $L_1(\theta)$ to denote the ARL of the first scheme with decision interval h_1 and reference value k_1, $L_2(\theta)$ for the second and $L_c(\theta)$ for the ARL of the combined schemes. Table 7.22 gives the cusum envelopes and values of $L_c(\theta)$ for a variety of snub-nosed V-masks by obtained computerized simulation

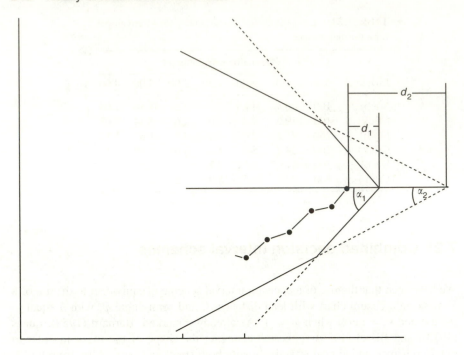

Fig. 7.10   Snub-nosed V mask.

and use of a reliable NAG random number generator. In control run lengths were based on 10^4 runs and off target ones on 10^3 runs. It shows that the ARL profile of each of the combined schemes lie above but very close to the cusum envelope for values of θ in the range 0.5σ to 4σ. The table clearly illustrates that a properly designed combination of two cusum schemes will produce schemes which from a practical point of view are very nearly uniformly best. From this viewpoint their profiles are so close to the envelope that there seems little point in trying to improve upon them.

Design of snub-nosed masks in practice

It is all very well to establish that the combination of two cusum schemes yield schemes with profiles close to the best which can be achieved, how in practice can we design them with specified values for $L_c(\theta)$? Can we devise expressions which permit us to do so with the aid say of a pocket calculator? The use of equations like (7.72), (7.73), and (7.74) together with the values given in Table 7.23 and a number of other simulations reported by Malihe Akhavan-Abdollahran indicate that this can indeed be achieved quite easily. Table 7.23 gives the profiles of the separate cusum schemes in the combinations of Table 7.22. From this table we see that for deviations from target ≥ 0.50 very close approximations to the profiles of the combined schemes are obtained by taking the lowest ARL of each separate scheme.

Table 7.22   Run length profiles of some snub-nosed V-masks

h_1 k_1	h_2 k_2		Deviation from target				
			0.00	0.50	1.00	2.00	3.00
1.17, 1.60	3.70, 0.50						
	$C(\mathbf{h}; \mathbf{k})$	81		21.4	7.3	2.7	1.5
	Envelope	81		18.1	7.0	2.5	1.4
1.32, 1.50	4.00, 0.50						
	$C(\mathbf{h}; \mathbf{k})$	100		24.0	7.9	2.8	1.55
	Envelope	100		19.3	7.4	2.6	1.44
1.80, 1.50	4.00, 0.50						
	$C(\mathbf{h}; \mathbf{k})$	150		25.8	8.3	3.1	1.81
	Envelope	150		22.2	8.2	2.8	1.53
1.78, 1.40	4.50, 0.50						
	$C(\mathbf{h}; \mathbf{k})$	200		30.1	9.0	3.2	1.73
	Envelope	200		24.2	8.7	3.0	1.59
1.95, 1.35	4.75, 1.35						
	$C(\mathbf{h}; \mathbf{k})$	250		33.5	9.5	3.3	1.77
	Envelope	250		25.8	9.2	3.1	1.64
2.05, 1.30	5.00, 0.50						
	$C(\mathbf{h}; \mathbf{k})$	300		36.1	10.1	3.3	1.83
	Envelope	300		27.2	9.5	3.2	1.68
2.20, 1.30	5.00, 0.50						
	$C(\mathbf{h}; \mathbf{k})$	350		36.6	10.2	3.4	1.92
	Envelope	350		28.3	9.8	3.2	1.71

Table 7.23   Profiles of the separate schemes in Table 7.22 computed from eqns (7.72) to (7.74)

	Deviation from target			
Scheme	0.00	0.50	1.00	2.00
C(1.17;1.60)	128.0	50.6	16.3	3.18
C(1.32;1.50)	152.1	54.1	16.3	3.18
C(1.78;1.40)	407.6	101.0	22.5	3.65
C(1.80;1.50)	646.0	154.5	30.2	4.10
C(2.05;1.30)	552.9	112.4	22.2	3.66
C(2.20;1.30)	817.4	145.5	25.3	3.88
C(3.70;0.50)	123.2	21.3	7.6	4.39
C(4.00;0.50)	168.0	24.2	8.2	4.69
C(4.50;0.50)	280.0	30.4	9.2	5.19
C(5.00;0.50)	465.9	36.5	10.3	5.69

Table 7.24   Values of $L_c'(0)$ given by eqns (7.72) to (7.75)

Scheme	$L_c(0)$	$L_c'(0)$
$C_2(1.17,3.70;1.60,0.50)$	81	79
$C_2(1.32,4.00;1.50,0.50)$	100	102
$C_2(1.80,4.00;1.50,0.50)$	150	152
$C_2(1.78,4.50;1.40;0.50)$	200	203
$C_2(2.05,5.00;1.30,0.50)$	300	296
$C_2(2.20,5.00;1.30,0.50)$	350	357

In addition, when testing is on target, values in the tables indicate that we can use an equation of the form

$$[\alpha/L_1(0)] + [\beta/L_2(0)] = 1/L_c'(0) \tag{7.97}$$

to obtain the ARL at AQL of snub-nosed masks. The procedure is therefore to use (7.72)–(7.74) to obtain $L_1(0)$ and $L_2(0)$ and (7.75) with $\alpha = 0.64$ and $\beta = 0.94$ for values of $L_c(0) \le 200$, and $\alpha = 0.99$ and β equal to 0.74 for $200 < L_c(0) \le 350$. Table 7.24 gives the values of $L_c'(0)$ when this procedure is used. From it we see that $L_c'(0)$ only differs marginally from $L_c(0)$.

Accordingly we find that we can devise a simple method to design schemes with a specified on target run length using the information obtained from simulations together with equations like (7.72) to (7.74). The method gives run lengths which are sufficiently close to the actual values to justify its use in practice. Run length values in Table 7.23 indicate that h_1, h_2, and k_1 for a specific $L_c(0)$ can be obtained using linear interpolation whilst $k_2 = 0.50$.

Suppose we require $L_c(0) = 150$, if we use the procedure just described h_1, would be halfway between 1.32 and 1.78, namely 1.55. For h_2, k_1, and k_2 we have $h_2 = 4.25$, $k_1 = 1.35$ and $k_2 = 0.50$. The snub-nosed scheme is therefore $C_2(1.55, 4.25; 1.45, 0.50)$. Estimates of the profile of this and other schemes given by eqns (7.72) to (7.75) shown in Table 7.25 confirm that they will give values for the on target ARL very close to its true value and have profiles close to the cusum envelope. This example also illustrates another important feature of combined schemes. Notice that the profile of the scheme $C_2(1.55, 4.25; 1.45, 0.50)$ given by $L_c'(0)$ is almost the same as the scheme $C_2(1.80, 4.00; 1.50, 0.50)$ of Table 7.22. This implies that there exist a whole variety of combined schemes with profiles close to one another which could be used in practice. It also implies, that having reached these last conclusions from simulation studies and the calculations described, there is little point conducting further expensive complex and time consuming simulation procedures to obtain schemes with features which are only marginally better than those just described.

We conclude from the comparisons discussed here that a properly designed snub-nosed V-mask goes a long way towards producing schemes which in practice are nearly uniformly best. We also conclude that equations like (7.72) to (7.75) give

Table 7.25 Values of $L_c'(0)$ given by equations (7.72) to (7.75)

	Deviation from target			
Scheme	0.00	0.50	1.00	2.00
$C_2(1.55,4.25;1.45,0.50)$	146	30.6	8.9	3.38
$C_2(1.92,4.75;1.35,0.50)$	245	36.7	9.9	3.62
$C_2(2.12,5.00;1.30,0.50)$	324	39.9	10.4	3.73

Table 7.26 Run Length profiles of the schemes of Bissell and Lucas

Scheme	0.00	0.50	1.00	2.00
Bissell	232	36	9.8	3.3
$C_2(1.89,4.66;1.4,0.5)$	232	32	9.3	3.3
$C_2(1.87,4.66;1.37,0.5)$	228	32	9.2	3.2
Lucas	320	54	10.6	3.4
$C_2(2.12,5;1.3,0.5)$	320	35	9.9	3.5
Envelope	320	28	9.0	3.2

estimates of $L(0)$ sufficiently close to its true value to justify their use in practical applications.

In view of this remark it is interesting to compare the profiles of appropriate snub-nosed masks with those of the semi-parabolic mask of Bissell and that of Lucas.

Using Bissell's mask is equivalent to running five decision interval schemes simultaneously. It is the scheme $C_5(\mathbf{h}, \mathbf{k})$ with $\mathbf{h} = (1.4, 1.85, 2.6, 3.65, 5)$, the corresponding values for k being $(1.7, 1.4, 1.1, 0.8, 0.5)$. Two simulations carried out by Akharan-Abdallaran gave a mean value for $L(0)$ of this scheme to be 232 which agrees with the value given by Bissell himself. Table 7.26 shows the profile of this scheme and that of $C_2(1.89, 4.66; 1.40, 0.50)$. The double-sided scheme obtained interpolating values of h and k in Table 7.22 is $C_2(1.87, 4.66; 1.37, 0.50)$. The profiles of this scheme estimated from eqns (7.72) to (7.75) are also given.

The profile of a Lucas mask with $L(0) = 320$ is also given in Table 7.26 and compared with that of the snub-nosed mask with equivalent on target ARL. In reality neither mask can claim to be superior to an appropriate snub-nosed one.

To illustrate the use of eqns (7.72) to (7.75) together with (7.97) consider the following clinical example.

An internal quality control scheme was required for oestradiol assays. To set it up, data was collected from the results of a number of consecutive batches of high quality control plasma pool. This was used to estimate the target mean m and the test standard deviation σ. The value of m was found to be 60 units whilst σ was 7.1 units. An assay is judged out of control when test results lie outside $m \pm \kappa\sigma$. The values of κ vary between 1.5 to 2.0. In this case it was 2.0. Clinical considerations required an ARL, of 120 days when testing is on target and 2 days when testing is off target to the extent of $\pm 2\sigma$. In addition, it was necessary to devise a scheme which detected any change from target as quickly as possible. To achieve these objectives it was decided to run two control tests a day and design a snub-nosed V-mask or equivalently consecutively operate two double-sided cusum decision interval schemes. This requires values of $h_1 k_1$, and $h_2 k_2$ which have an on-target ARL of 240 and a value close to 4 when testing is off target to the extent of $\pm 2\sigma$.

Solution

The collected data indicated that the test standard deviation of 7.1 units did not change with movement off target and that test results could be taken to be normally distributed. For convenience let us consider the situation where the test results are standardized to have unit standard deviation. Equation (7.75) indicates that for this data $h_1 = 1.95$ and $k_1 = 1.35$ will give an ARL close to 4 for off target values of 2. This equation gives an average run length equal to 3.7 whilst

$$L_a(0; 0.65) = (h/0.65) + 0.72 + 0.18 \exp(-1.3)$$

gives $L_a(0; 0.65)$ when $h_1 = 1.95$ equal to 3.72. When testing is on target θ in eqn (7.72) is -1.35 and

$$L_a(0; -1.35) = 5.4821 \exp(2.7h) - 1.3297 - h/(1.35).$$

For $h = 1.95$ $L_a(0; -1.35) = 1058$ for the operation of a single-sided decision scheme, so that from eqn (7.9) the ARL for a two-sided scheme is 529. To obtain values for h_2 and k_2 note from Table 7.22 that k_2 is constant and equal to 0.50. From eqn (7.97)

$$[0.99/529 + 0.74/L_2(0)]^{-1} = 240,$$

so that $L_a(0; -0.50)$ needs to be 320. Substituting $\theta = -0.50$ into eqn (7.72) gives

$$L_a(0; -0.50) = 6.373 \exp(h) - 3.989 - 2h.$$

The value of h which gives $L_a(0; -0.50) = 640$ is 4.63, so that the combination we require for standardized values is the two-sided scheme $C(1.95, 1.35; 4.63, 0, 50)$.

We have seen that the best two-sided single cusum decision interval scheme with ARLs close to 240 and 4 has, in this case, central reference value $k = 1$.

Table 7.27   Profiles of $C(1.95, 1.35; 4.63.0.50)$ and $C(2.3, 1.0)$ schemes and Rule 2 with action and warning lines 4.91 and 1.68 standard deviations away from target

Deviation from target	$C(1.95.1.35; 4.43.0.50)$	$C(2.3.1.0)$	Rule 2
0.00	239	241	243
0.20	116	165	175
0.50	35.5	57	67.5
1.00	9.62	12.1	14.3
2.00	3.72	3.00	3.48
3.00	1.88	2.00	2.16

Table 7.28   Values of $L(0)$ for different h and θ for X_i being $N(\theta; 1)$ and θ negative

				h			
$-\theta$	2.0	2.5	3.0	3.5	4.0	4.5	5.0
1.30	958						
1.20	614						
1.10	396						
1.00	259	716					
0.90	171	429					
0.80	115	262	585				
0.70		163	327	680			
0.60	54	104	198	361	667	1219	
0.55					467	817	
0.50		68	100	200	338	560	930
0.45						387	606
0.40	28		74	115	177	272	413
0.30		32		70	104	142	198
0.20			33	45	60	80	104
0.10		17.4		30	41	49	60
0.00	10	13.4	17.1	21.7	26.1	32	38

Equation (7.72) is then

$$L_a(0; -1.0) = 4.89 \exp(2h) - 1.76 - h$$

and $h = 2.3$ has $L_a(0; -1.0) = 480$.

The profiles of both of these schemes is indicated in Table 7.27. The profile of a two-sided rule with warning and action lines and equivalent ARLs is also included in the table. The rule used was a Rule 2. The sample size had to be increased by

Table 7.29   Values of $L(0)$ for different h and θ for X_i being $N(\theta; 1)$ and θ positive

θ	2.0	2.5	3.0	3.5	4.0	4.5	5.0
0.1	8.2	10.7					
0.2	6.9	8.7	10.2	12.8			
0.3			8.7	10.4	12.6	13.5	15.2
0.4	5.0	6.2	7.5	8.7	10.6	11.1	12.3
0.5					8.7	9.4	10.4
0.6	4.0	4.8	5.6	6.4	7.3	8.1	8.9
0.8	3.2	3.9	4.5	5.1	5.8	6.4	7.0
1.0	2.7	3.3	3.7	4.3	4.9	5.2	5.7
1.2	2.4						
1.3		2.6	3.0	3.4			
1.5	2.0	2.3	2.7	3.0	3.6	4.2	
2.5					2.6	3.7	3.0
3.5							2.4

20 per cent to obtain values close to 240 and 4. Thus, if the sample size needed for the above cusum schemes was n, that of the appropriate two-point scheme was $1.2n$. For the standardized test results the action lines are placed 4.91 units away from the target value whilst the warning lines are a distance of 1.68 units away from it. The table clearly indicates the superiority of the $C(1.95, 1.35; 4.63, 0.50)$ scheme.

Appendix Tables

Appendix Table I
Random numbers between 0 and 100

97	92	55	88	67	23	94	13	4	83	97	3
33	36	49	28	3	60	32	46	87	25	71	63
42	20	33	64	64	81	72	40	10	42	63	18
5	35	44	16	94	9	68	79	46	63	91	55
8	88	45	67	89	40	18	17	62	38	76	93
53	6	53	23	66	76	100	21	37	55	39	55
7	21	63	54	74	50	59	51	87	95	52	80
6	83	33	22	50	14	21	59	38	77	9	7
37	44	61	36	68	56	18	15	42	41	35	62
4	69	58	45	34	95	11	70	80	51	86	94
11	64	72	82	97	41	45	93	82	59	63	4
45	38	53	57	72	22	70	8	78	63	39	61
44	78	96	28	66	23	31	61	64	75	57	35
3	96	18	54	23	69	38	41	13	27	43	82
40	75	33	81	95	36	6	84	39	100	58	94
30	77	11	26	95	100	20	12	91	74	35	16
6	65	94	0	21	48	81	79	5	1	55	10
18	54	79	93	1	71	93	42	75	83	45	29
73	45	19	3	60	72	88	4	78	65	43	44
71	61	66	52	18	36	29	95	21	92	13	8
60	89	14	42	75	30	53	46	42	30	24	24
29	61	56	17	54	20	21	27	17	1	64	30
57	73	24	15	2	56	83	61	87	67	90	18
36	26	30	55	49	12	20	78	42	57	71	41
36	18	85	67	22	54	32	64	72	71	72	85
48	93	48	38	62	55	97	96	4	35	42	17
58	34	93	68	27	91	94	26	1	33	70	24
16	3	20	60	62	27	97	32	15	84	76	45
100	93	74	2	8	13	36	73	69	41	77	13
40	81	0	33	54	72	92	1	56	20	53	32
83	7	2	27	69	36	56	66	82	31	98	44
79	82	82	94	90	72	12	9	31	45	14	6
11	39	71	7	94	8	81	91	34	87	34	50
4	37	98	90	17	20	52	44	47	12	82	49

Appendix Table II
Values of $\Phi(x)$

x	0.00	0.01	0.02	0.03	0.04	0.05	0.06	0.07	0.08	0.09
0.0	.5000	.5040	.5080	.5120	.5159	.5199	.5239	.5279	.5319	.5359
0.1	.5398	.5438	.5478	.5517	.5557	.5596	.5636	.5675	.5714	.5753
0.2	.5793	.5832	.5871	.5910	.5948	.5987	.6026	.6064	.6103	.6141
0.3	.6179	.6217	.6255	.6293	.6331	.6368	.6406	.6443	.6480	.6517
0.4	.6554	.6591	.6628	.6664	.6700	.6736	.6722	.6808	.6844	.6879
0.5	.6915	.6950	.6985	.7019	.7054	.7088	.7123	.7157	.7190	.7224
0.6	.7257	.7291	.7324	.7357	.7389	.7422	.7454	.7486	.7518	.7549
0.7	.7580	.7611	.7642	.7673	.7704	.7734	.7764	.7794	.7823	.7852
0.8	.7881	.7910	.7939	.7967	.7995	.8023	.8051	.8078	.8106	.8133
0.9	.8159	.8186	.8212	.8238	.8264	.8289	.8315	.8340	.8365	.8389
1.0	.8413	.8438	.8461	.8485	.8508	.8531	.8554	.8577	.8599	.8621
1.1	.8643	.8665	.8686	.8708	.8729	.8749	.8770	.8790	.8810	.8830
1.2	.8849	.8869	.8888	.8907	.8925	.8944	.8962	.8980	.8997	.9015
1.3	.9032	.9049	.9066	.9082	.9099	.9115	.9131	.9147	.9162	.9177
1.4	.9192	.9207	.9222	.9236	.9251	.9265	.9279	.9292	.9306	.9319
1.5	.9332	.9345	.9357	.9370	.9382	.9394	.9406	.9418	.9430	.9441
1.6	.9452	.9463	.9474	.9485	.9495	.9505	.9515	.9525	.9535	.9545
1.7	.9554	.9564	.9573	.9582	.9591	.9599	.9608	.9616	.9625	.9633
1.8	.9641	.9649	.9656	.9664	.9671	.9678	.9686	.9693	.9699	.9706
1.9	.9713	.9719	.9726	.9732	.9738	.9744	.9750	.9756	.9762	.9767
2.0	.9772	.9778	.9783	.9788	.9793	.9798	.9803	.9808	.9812	.9817
2.1	.9821	.9826	.9830	.9834	.9838	.9842	.9846	.9850	.9854	.9857
2.2	.9861	.9865	.9868	.9871	.9875	.9878	.9881	.9884	.9887	.9890
2.3	.9893	.9896	.9898	.9901	.9904	.9906	.9909	.9911	.9913	.9916
2.4	.9918	.9920	.9922	.9925	.9927	.9929	.9931	.9932	.9934	.9936
2.5	.9938	.9940	.9941	.9943	.9945	.9946	.9948	.9949	.9951	.9952
2.6	.9953	.9955	.9956	.9957	.9959	.9960	.9961	.9962	.9963	.9964
2.7	.9965	.9966	.9967	.9968	.9969	.9970	.9971	.9972	.9973	.9974
2.8	.9974	.9975	.9976	.9977	.9977	.9978	.9979	.9980	.9980	.9981
2.9	.9981	.9982	.9983	.9983	.9984	.9984	.9985	.9985	.9986	.9986
3.0	.99865	.9987	.9987	.9988	.9988	.9989	.9989	.9989	.9990	.9990
3.1	.9990	.9991	.9991	.9991	.9992	.9992	.9992	.9992	.9993	.9993
3.2	.9993	.9993	.9994	.9994	.9994	.9994	.9994	.9995	.9995	.9995

Appendix Table III

χ^2 values for which the probability $Pr(\chi p^2; v) \geq p$ on v degrees of freedom

p \ v	0.005	0.01	0.025	0.05	0.1	0.9	0.95	0.975	0.99	0.995
1	0.393 (4)	0.157 (3)	0.983 (3)	0.393 (2)	0.158 (1)	2.706	3.841	5.024	6.635	7.879
2	0.010	0.201 (1)	0.506 (1)	0.103	0.211	4.605	5.991	7.378	9.210	10.597
3	0.717 (1)	0.115	0.216	0.352	0.584	6.251	7.815	9.348	11.345	12.838
4	0.207	0.297	0.484	0.711	1.064	7.779	9.488	11.143	13.277	14.860
5	0.412	0.554	0.881	1.145	1.610	9.236	11.070	12.833	15.086	16.750
6	0.676	0.872	1.237	1.635	2.204	10.645	12.592	14.449	16.812	18.548
7	0.989	1.239	1.690	2.167	2.833	12.017	14.067	16.013	18.475	20.278
8	1.344	1.646	2.180	2.733	3.490	13.362	15.507	17.535	20.096	21.955
9	1.735	2.088	2.700	3.325	4.168	14.684	16.919	19.023	21.666	23.589
10	2.156	2.558	3.247	3.940	4.865	15.987	18.307	20.483	23.209	25.188
11	2.603	3.053	3.816	4.575	5.578	17.275	19.675	21.920	24.725	26.757
12	3.074	3.571	4.404	5.226	6.304	18.549	21.026	23.337	26.217	28.300
13	3.565	4.107	5.009	5.892	7.042	19.812	22.362	24.736	27.688	29.820
14	4.075	4.660	5.629	6.571	7.790	21.064	23.685	26.119	29.141	31.319
15	4.601	5.229	6.262	7.261	8.547	22.307	24.996	27.488	30.578	32.801
16	5.142	5.812	6.908	7.962	9.312	23.542	26.296	28.845	32.000	34.267
17	5.697	6.408	7.564	8.672	10.085	24.769	27.587	30.191	33.409	35.718
18	6.265	7.015	8.231	9.390	10.865	25.989	28.869	31.526	34.805	37.156
19	6.844	7.633	8.907	10.117	11.651	27.204	30.144	32.852	36.191	38.582
20	7.434	8.260	9.591	10.851	12.443	28.412	31.410	34.170	37.566	39.997
21	8.034	8.897	10.283	11.591	13.240	29.615	32.671	35.479	38.932	41.401
22	8.643	9.542	10.982	12.338	14.041	30.813	33.924	36.781	40.289	42.796
23	9.260	10.196	11.698	13.091	14.848	32.007	35.172	38.076	41.638	44.181
24	9.886	10.856	12.401	13.848	15.659	33.196	36.415	39.364	42.980	45.559
25	10.520	11.524	13.120	14.611	16.473	34.382	37.653	40.646	44.314	46.928
26	11.160	12.198	13.844	15.379	17.292	35.563	38.885	41.923	45.642	48.290
27	11.808	12.879	14.573	16.151	18.114	36.741	40.113	43.195	46.963	49.645
28	12.461	13.565	15.308	16.928	18.939	37.916	41.337	44.461	48.278	50.993

Bibliography

Barnard, G.A. (1959). Control charts and stochastic processes. *J. R. Statist. Soc.* B, **21**, 239–57.

Barnard, G.A. (1954). Sampling inspection and statistical decisions. *J. R. Statist. Soc.* B, **16**, 151–65.

Bissell, A.F. (1978). An attempt to unify the theory of quality control procedures. *Bias*, **5**(2) 113–28.

Bissell, A.F. (1979). A semi parabolic-mask for cusum charts. *Statistician*, **28**, 1–7.

Bissell, A.F. (1969). Cusum techniques for quality control. *Appl. Statist.*, **18**, 1–30.

Box, G.E.P., Bisgaard, S. and Fung. (1990). Designing industrial experiments. B.B.F. Books, Illinois.

Burman, J.P. (1946). Sequential sampling formulae for a binomial population. *J. R. Statist.*, J. R. Statist. Soc. Sup., **8**, 98–103.

Butner, J., Borth, R., Broughton, P.M. and Bowyer, R.C. (1980). Quality control in clinical chemistry: Part 4, *Clinica Chemica Acta.* **106**, 109F–20F.

Blacksell, S.D., Gleeson, L.J., Lunt R. and Chamnanpood, C. (1994). Use of combined Shewhart–cusum control charts in internal quality control of enzyme-linked immunosorbent assays for the typing of foot and mouth disease virus antigen. *Revue Scientifique et Technique*, **13**(3), 687.

Backsell, S.D. *et al.* (1996). Implementation of internal laboratory quality control procedures for the monitoring of ELISA performance at a regional veterinary laboratory. *Veterinary Microbiology*, **51**(1–2), 1–9.

Carter, A.H. (1944). Approximation to percentage points of the z-distribution. *Biometrika*, **34**, 352–8.

Chang, S.I. and Samual, T.R. (1998). A control point methodology for cusum control charts. *Computers and Industrial Engineering*, **34**(3), 565.

Chin, W.K. (1974). The economic design of cusum charts for controlling normal means. *Appl. Statist.*, **23**, 420–3.

Chin, W.K. and Lenny, M.P.Y. (1980). A new Baysian approach to quality control charts. *Metrika*, 243–53.

Chin, W.K. and Weatherall, G.B. (1973). The economic design of continuous inspection procedures: a review paper. *Int. Statist. Rev.*, **41**, 357–73.

Cembrowski, G.S., Westgard, J.O., Eggert, A.A. and Torem, Jr, E.C. (1975). Trend detection in control data: optimisation and interpretation of Triggs technique for trend analysis. *Clinical Chemistry*, **21**, 1396–405.

Cox, D.R. (1949). The use of range in sequential analysis. *J. R. Statist.*, **11**, 101–14.

Data analysis and quality control using cusum techniques, (1980). Part 2, Decision rules and statistical tests for cusum charts and tabulations. B.S. 5703. British Standards Institution.

Deming, W.E. (1986). *Out of the crisis.* Cambridge University Press.

Douglas, A.E. (1914). Photographic periodogram of sun-spot numbers. *Astro. Phys. J.*, **40**, 326.

Douglas, A.E. (1915). An optical periodograph. *Astro. Phys. J.*, **41**, 173.

Dudding, B.P. and Jennet, W.J. (1942). Quality control charts. British Standard 600R. British Standard Institution, London.

Duncan, A.J. (1965). Quality control and industrial statistics. Richard D Irwin Inc., Hamewood, Illinois.

Dunstan, F.D.J., Nix, A.B.J. and Reynolds, J.F. (1979). *Statistical tables.* RND Publications.

Edwards, R.W. (1980). Internal quality control using the cusum chart and truncated V-mask procedure. *Ann. Clin. Biochem.*, **17**, 205–11.

Ewan, W.D. and Kemp, K.W. (1960). Sampling inspection of continuous processes with no auto-correlation between successive results. *Biometrika*, **47**, 363–80.

Foster, G.N.R. (1946). Some instruments for the analysis of time series and their application to textile research. *J. R. Statist. Soc.* B, **8**, 42.

Girshick, M.A. and Rubin, H. (1952). A Bayes approach to a quality control model. *Ann. Math. Statist.*, **23**, 114–28.

Garland, S.W. Lees, B. and Stevenson, J.C. (1997). DXA Longtidunal quality control: a comparison of inbuilt quality assurance, visual inspection multi rule Shewhart charts and cusum analysis. *Osteoporosis Int.* **7**(3), 231.

Goel, A.L. and Wu, S.M. (1971). Determination of ARL and a contour nomogram for cusum charts to control normal mean. *Technometrics*, **13**(2), 221–30.

Kemp, K.W. (1957). An optical periodograph for use in the analysis of continuous measurements. *Appl. Statist.*, **6**(3), 208–13.

Kemp, K.W. (1958). Formulae for calculating the operating characteristic and average sample number of some sequential tests. *J. R. Statist. Soc.* B, **20**, 379–86.

Kemp, K.W. (1961). The average run length of the cumulative sum chart when a V-mask is used. *J. R. Statist. Soc.* B, **23**, 149–53.

Kemp, K.W. (1962). The use of cumulative sums for sampling inspection schemes. *Appl. Statist.*, **11**, 16–31.

Kemp, K.W. (1967). Formal expressions which can be used for determination of the operating characteristics and average sample number of a simple sequential test. *J. R. Statist. Soc.* B, **29**(2), 248–62.

Kemp, K.W. (1970). Convergent sequences of bounds for some parameters of a simple sequential test. *J. R. Statist. Soc.* B, **32**(2), 241–53.

Kemp, K.W. (1971). Formal expressions which can be applied to cusum charts. *J. R. Statist. Soc.* B, **33**(3), 331–60.

Kemp, K.W., Nix, A.B.J., Wilson, D.W. and Griffiths, K. (1978). Internal quality control for radioimmunoassays. *J. Endocrinol.*, **76**, 203–10.

Kemp, K.W., Nix, A.B.J., Wilson, D.W. and Griffiths, K. (1979). *Quality control in clinical endocrinology.* Eighth Tenovus Workshop. 1979. Alpha Omega Publishing Ltd.

Knight, A.C. and Williams, E.D. (1992). An evaluation of cusum analysis and control charts applied to quantitative gamma-camera uniformity parameters for automated quality control. *Euro. J. Nucl. Med.*, **19**(2), 125–30.

Levey, S. and Jennings, E.R. (1990). The use of control charts in the clinical laboratory. *Am. J. Clin. Pathol.*, **20**, 1059–66.

Lordin, G. (1971). Procedures for reacting to a change in distribution. *Ann. Math. Statist.*, **42**(6) 1897–908.

Lucas, J.M. (1973). A modified V-mask control scheme. *Technometrics*, **15**, 833–47.

Lucas, J.M. (1982). Combined Shewhart–cusum quality control schemes. *J. Quality Technol.*, **14**, 51–9.

Lyle, P. (1954). The construction of nomograms for use in statistics. *Appl. Statist.*, Part 1, **3**, 116–25.

Magari, Y., Sato, M., Toshimitu, S., Toujinbara, M., Kumagaya, Y., Taguchi I. *et al.* (1996). Standardization of serum lipid examination: an attempt in Oita. 1996. *Jap. J. Clin. Pathol.*, **44**(11), 1043–9.

Malihe, Akhavon-Abdollahian (1982). Optimal continuous inspection schemes. Ph.D. thesis, University of Wales.

Mosteller, F. (1941). Note on the application of runs to control charts. *Ann. Math. Statist.*, **32**, 232.

Nix, A.B. and Bishop J. (1993). Comparison of quality control rules used in clinical chemistry. Clinical Chemistry, **39**(8), 1638–49.

Page, E.S. (1954). Continuous inspection schemes. *Biometrika*, **41**, 100–15.

Page, E.S. (1957). On problems in which a change in parameter occurs at an unknown point. *Biometrika*, 248–56.

Page, E.S. (1955). Control charts with warning lines. *Biometrika*, **42**, 243–54.

Pearson, D. and Cawte, S.A. (1997). Long-term quality control of DXA: a comparison of Shewhart rules and cusum charts. *Osteoporosis Int.*, **7**(4), 338–43.

Pearson, E.S. and Hartley, H.O. (0000). *Biometrika tables for statisticians*, Vol. 1. Cambridge University Press.

Pearson, E.S. (1941–2). The probability integral of the range in samples of n observations from a normal population. *Biometrika*, **32**, 301–8.

Regula, D.A. (1976). Optimal cusum procedures to detect a change in distribution for a gamma family. Ph. D. thesis, Case Western Reserve University.

Roberts, S.W. (1959). Control charts based on geometric moving averages. *Technometrics*, **1**, 239–50.

Roberts, S.W. (1966). A comparison of some control chart procedures. *Statistician*, **8**(3), 411–30.

Robinson, P.B. (1978). Average run length of geometric moving average charts by numerical methods. *Technometrics*, **20**(1), 85–93.

Rowlands, R.J., Nix, A.B.J., Abdollahian, M.A. and Kemp, K.W. (1982). Snub nosed V-mask control schemes. *Statistician*, **31**, 133–42.

Rowlands, R.J. (1976). Formula for performance characteristics of cumulative sum schemes when the observations are autocorrelated. PhD thesis, University of Wales.

Rowlands, R.J. and Nix, A.B.J. (1981). Optimal continuous inspection schemes. *Meth. Operational Res.*, **41**, 201–4.

Rowlands, R.J., Griffiths, K., Kemp, K.W., Nix, A.B.J., Richards, G. and Wilson, D.W. (1983). Applications of cusum techniques to routine monitoring of analytical performance in clinical laboratories *Statistics in Medicine*, **2**, 141–43.

Rowlands, R.J., Wilson, D.W., Nix, A.B.J., Kemp, K.W. and Griffiths, K. (1980). Advantages of cusum techniques for quality control in clinical chemistry. *Clinica Chemica Acta*, **108**, 393–7.

Searle, G.C.L. (1974). A sequential test with power one and its application to cusum charts. Ph. D. thesis, University of New South Wales.

Shewhart, W.A. (1931). *Economic control of quality of manufactured product*. Van Nostrand, New York.

Siegmund, D. (1975). Error probabilities and average sample number of the sequential probability ratio test. *J. R. Statist. Soc.* B, **37**(3), 394–401.

Steiner, S.H., Cooke, R. J. and Farewell, V.T. (1999). Monitoring paired binary surgical outcomes using cumulative sum charts. *Statistics in Medicine*, **18**(1), 69–86.

Swed, S. and Eisenhart, C. (1943). Tables for testing Randomness of sampling in a sequence of alternatives. *Ann. Math. Statist.*, **XIV**, 66–87.

Smithies, F. (1958). *Integral equations*. Cambridge Tracts in Mathematics and Mathematical Physics. No 49. Cambridge University Press.

Taguchi, G. (1987). *System of experimental design*. UNIPUB, New York. Kraus International Publications.

Tables of the bivariate normal distribution function and related functions. US Dept of Commerce, National Bureau of Standards Applied Mathematics Series.

Van Dobben de Brun, C.S. (1968). Cumulative sum techniques. Griffin, London.

Van Rij, A.M., Macdonald, J.R., Pettigrew, R.A., Putterill, M.J., Reddy, C. K. and Wright, J.J. (1995). Cusum as an aid to early assessment to the surgical trainee. *Br. J. Surgery*, **82**(11), 1500–3.

Wald, A. (1944). On the cumulative sums of random variables. *Ann. Math. Statist.*, **15**, 283–96.

Wald, A. (1945). Sequential tests of statistical hypotheses. *Ann. Math. Statist.*, **16**, 117–86.

Wald, A. (1947). *Sequential analysis*. Wiley, New York.

Westgard, J.O. *et al.* (1977). Combined Shewhart–cusum chart for improved quality control in clinical chemistry. *Clin. Chem.*, **23**, 1881–7.

Westgard, J.O., *et al* (1977). Performance characteristics of rules for internal quality control: Probabilities for false rejection and error detection. *Clin. Chem.*, 1857–67.

Westgard, J., Barry, P.L. and Hunt, M.R. (1981). A multi-rule Shewhart chart for quality control in clinical chemistry. *Clin. Chem.*, **27**, 493–501.

Wetherill, G.B. (1977). *Sampling inspection and quality control*. Chapman and Hall, London.

Wilson, P.W., Griffiths, K., Kemp, K.W., Nix, A.B.J. and Rowlands, R.J. (1979). Internal quality control of radioimmunoassays: monitoring error. *J. Endocrinol.*, **80**, 365–72.

Index